CW00724257

Chimpanzee Behavior in the Wild

Preface

Where We Stand

Field workers—scientists of animal (including human!) behavior in nature—have long been fascinated by wild chimpanzees. A person who once has studied wild chimpanzees will be eager to observe them again. A person who has studied them twice will make every effort to continue the study, unless prevented from doing so. In short, behavioral primatology is addictive!

Many people, among them Jane Goodall, Richard Wrangham, and I, do not regret that they have dedicated their whole lives to the study of wild chimpanzees. This is because the apes' behavior is always challenging: chimpanzees are cheerful, charming, playful, curious, beautiful, easygoing, generous, tolerant, and trustworthy most of the time, but also are cautious, cunning, ugly, violent, ferocious, bloodthirsty, greedy, and disloyal at other times. We human beings share both the light and dark sides with our closest living relatives.

For decades, we have documented huge across-population variation in behavior, as well as within-population variation. Cultural biology (now called cultural primatology), as proposed 60 years ago by Kinji Imanishi, recently has flourished. However, until now there has been no extensive glossary with illustrations, upon which systematic behavioral comparison could be based. In addition to fellow field workers, we hope that zoo keepers, laboratory researchers and technicians, veterinarians, conservationists, eco-tourism managers, educators and students now will be able to widen and deepen the scope of their activities by making use of the descriptions and audio-visual images in this volume. We hope that laypersons also will find this an enjoyable read at home on the sofa!

The team of authors began to use videography in 1999 and launched this editing project in 2003, and we have invested most of our time in it over the last 3 years. After much time and effort spent, the result is perhaps the first attempt to compile a comprehensive ethogram of a single species of mammal. This is a cooperative product in the truest sense of the word. If any of us had been absent, then we would not have been able to complete this work. Roles were allocated to film, copy, playback, select, and edit, and to edit, interpret, name, describe, reference, and synthesize. The resulting collaborative efforts have been discussed and refined 20 times over in the process.

We hope that this book and its accompanying DVD will help to conserve these irreplaceable creatures forever.

Toshisada Nishida

Contents

Color Plates

1. A juvenile female, Xantip, watches her mother, Xtina, removing a sand flea from her toe. (Photo by Agumi Inaba, Sept. 2007)
2. An adult male, Primus, ingests a mature leaf while chewing meat. (Photo by Agumi Inaba, Sept. 2007)
3. An adult female and two juvenile females licking a rock. (Photo by Agumi Inaba, Aug. 2007)
4. Grooming party of six adult males. (Photo by Agumi Inaba, Nov. 2007)
5. An adult female, Gwekulo, resting in the bush with her adopted juvenile daughter, Puffy. (Photo by Koichiro Zamma, Dec. 2006)
6. A juvenile female, Mitsue, looking at her own image in a stream. (Photo by Toshisada Nishida, Aug. 2007)
7. An adult male and two adult females grooming on a fallen log. (Photo by Agumi Inaba, Dec. 2007)
8. A newly-immigrated female, Ua, who did not descend to the ground when human observers were present. (Photo by Koichiro Zamma, Oct. 2006)
9. An adolescent male, Caesar, self-grooming. (Photo by Takahisa Matsusaka, Aug. 2009)
10. A juvenile female, Mitsue. (Photo by Takahisa Matsusaka, Aug. 2009)

T. Nishida et al., *Chimpanzee Behavior in the Wild: An Audio-Visual Encyclopedia*,
DOI 10.1007/978-4-431-53895-0_1, © Springer 2010

7

8

9

10

Introduction

This book is an *audio-visual ethogram* that lists, describes, and defines, with video clips, all the known behavioral patterns of wild chimpanzees. As far as we know, this is the first attempt to do this for any mammalian species on earth. The behavioral patterns listed have been recorded mostly for the M-group chimpanzees of the Mahale Mountains National Park, Tanzania, where we have been doing research since 1965. However, we have made a great effort to add patterns described from other study-sites known to us only through articles, books and videos. This book is based on a previous ethogram that had neither video clips nor illustrations (Nishida et al. 1999), but the glossary has been much expanded. Since 1999, videographic recording often has been used at Mahale for research on various topics, particularly social interactions, although earlier cinematographic records were occasionally made from 1971 onwards. We have compiled video clips for almost all of the behavioral patterns listed. From 1999 to 2009, new long-term study sites have been established (Stumpf 2007), and many articles and books dealing with the behavioral patterns of chimpanzees have appeared (e.g., Boesch and Boesch-Achermann 2000; Whiten et al. 2001; Boesch et al. 2002; McGrew 2004; Reynolds 2005; Schoening et al. 2008). Research on gestures in captivity has also been published (e.g. Call and Tomasello 2007a). With this new material, it is now easier to recognize possible differences across sites than was the case in 1999.

It is well-known that chimpanzees show much local, as well as age, sex, and idiosyncratic, variation in behavior (e.g., Goodall 1973; Nishida 1987; McGrew 1998, 2004; Whiten et al. 1999, 2001; Call and Tomasello 2007b). What implications does the compilation of the audio-visual ethogram have for these issues?

Our book's purpose is to stimulate and to facilitate ape behavioral studies by addressing the questions below. In particular, this publication pays attention to some behavioral variations that have been neglected so far, and we hope that this will encourage more detailed comparative and developmental studies.

First, we address the question of what kind of behavioral patterns may have been present in the last common ancestor (LCA) of *Pan* and *Homo*. This is a crucial question if we wish to reconstruct the behavior and ecology of early hominids such as the australopithecines, because they may have had similar patterns (Toth and Schick 2009; McGrew 2010).

T. Nishida et al., *Chimpanzee Behavior in the Wild: An Audio-Visual Encyclopedia*,
DOI 10.1007/978-4-431-53895-0_2, © Springer 2010

Second, it is often said that among animals only human beings depend much more on learning than on instinct. However, the extent to which chimpanzees depend on learning, especially social learning, is not well understood (Whiten et al. 2001). A simple behavioral pattern such as social scratch not only is absent at some sites (Nakamura et al. 1999) but also way of scratching varies from site to site in its details (Nishida et al. 2004). Audio-visual representations help to persuade skeptics of the richness of this variation.

Third, how do chimpanzees acquire behavioral patterns as they mature? In relation to the second question above, we study the development of behavioral patterns. Some behavioral patterns are learned from the mother, but others from playmates (e.g., Matsusaka et al. 2006), as with humans (Harris 1998). Only recently has it been shown that social customs such as the grooming-hand-clasp develop more slowly than do ant-fishing techniques (Nakamura and Nishida 2008). Audio-visual data make it easier to compare changes in the behavioral patterns of individuals as they mature.

Fourth, we study individuality and idiosyncracy in behavior. Little attention has been paid to this aspect of behavior, with a few notable exceptions (e.g., Goodall 1973; Kummer and Goodall 1985; Nishida 1994, 2003b). Idiosyncratic behavioral patterns are important because they are the seeds or origins of local culture (Kummer and Goodall 1985; Nishida et al. 2009).

Fifth, in which domains are local behavioral differences most remarkable? This should be addressed if we want to infer the evolutionary origins of human culture (Nishida et al. 1999). The extent of local differences has not been well-investigated, except for tool use (McGrew 1992, 2004), partly because most study-sites of chimpanzees lack a detailed ethogram. As new research sites have been opened, new behavioral patterns have been reported one after another, such as digging in the ground with implements (Hernandez-Aguilar et al. 2007) or penetrating it with perforating tools (Sanz et al. 2004). In order to know the most variable domains, however, we need for the full range of behavioral patterns to be illuminated, despite the continuing attractiveness of tool use. It may be that social behavior shows more local and individual differences than does tool using.

Sixth, heretofore unknown sex differences in behavioral patterns may emerge as more detailed observations are made. For example, through video-recording, we have noticed that immature females tend to play more in trees than do their male counterparts, and when females play on the ground, they tend to perform more somersaults than do males (Nishida, unpublished data).

Seventh, this book also may contribute to the further testing of the gestural origin hypothesis of human language origins (e.g. Hewes 1973; Corbalis 2002). Gestures appear to be integral to spoken language, because they accompany the speech of people who are blind from birth (Iverson and Goldin-Meadow 1998). Observations of spontaneous behavior in captivity show that chimpanzees and bonobos use brachio-manual gestures more flexibly across contexts than they do facial expressions and vocalizations. Gestures seem less closely tied to particular emotions, hence they possess more adaptable functional potential (Pollick and de Waal 2007). Pollick and de Waal (2007) advocate the gestural origins of language hypothesis because gestures are evolutionarily younger, and thus likely to be under greater cortical control than are facial or vocal signals. Few studies of hand preference in gestural communications have been done (e.g. Hopkins and Morris 1993), although studies of lateralization of tool use are many (e.g. Boesch 1991b; Sugiyama et al. 1993; McGrew and Marchant 1999; Matsuzawa et al. 2001; Marchant and McGrew 2007). Plentiful examples of gestural communications shown in this book might stimulate research on the symbolic connotations of ape gestures and on hand preference in gestures.

Last but not least, we expect that our efforts at exhaustive listing of behavioral patterns will merge in future with research on the chimpanzee genome. Genomic research on chimpanzees, although successful (Chimpanzee Sequencing & Analysis Consortium 2005), is still far from explaining the genetic basis of behavioral patterns. We anticipate that this ethogram will contribute to the synthesis of the chimpanzee genome study.

The audio-visual ethogram will be useful not only in facilitating more precise comparisons between sites, but also in providing students and ordinary people with knowledge and materials that will stimulate them to conserve great apes.

Listing the behavioral patterns of a local population of chimpanzees in principle includes: (1) behavioral patterns of the LCA of *Pan* and *Homo*; (2) patterns common to the genus *Pan*, namely chimpanzees and bonobos, but not to *Homo*; (3) patterns common to the chimpanzee *Pan troglodytes*, but not the bonobo *Pan paniscus*; (4) patterns common to eastern (*P. t. schweinfurthii*) and central (*P. t. troglodytes*), but not western (*P. t. verus*) and Nigerian (*P. t. vellerosus*) chimpanzees, in view of the recent cladistic analyses of culture (e.g. Lycett et al. 2007, 2009); (5) patterns unique to the eastern chimpanzees, *P. t. schweinfurthii*; (6) patterns unique to the Mahale population; (7) *customary* patterns unique to many individuals (at least to most members of an age or sex class) of M group chimpanzees; (8) *habitual* patterns limited to a few individuals of M group; (9) *rare* patterns observed only in a single individual (idiosyncracy).

The class of behavioral category outlined above to which each behavioral pattern belongs will not be conclusively clear until extensive research and comparison is done across many groups of local populations. However, a preliminary estimate can be made for some of the best-known patterns; for example, few would deny that 'groom' was in the behavioral repertoire of the LCA. In such cases, we put 'Category 1', in the column termed "Category" after each behavioral pattern's description.

Studies of the behavioral patterns of *Homo sapiens* (e.g., Eibl-Eibesfeldt 1972; McGrew 1972; Morris 1977, 1981, 1985) and bonobo (e.g., Kano 1980, 1992, 1998; Kuroda 1980, 1984; Susman et al. 1980; Susman 1984; de Waal 1988; Hohmann and Fruth 2003; Pika et al. 2005; Pika 2007; Pollick et al. 2008) have been published, although to a lesser extent than studies on chimpanzees. Therefore, behavioral comparisons between chimpanzee, bonobo and human can be made to some extent. Many authors have pointed out behavioral similarities and differences between human and chimpanzee (e.g., de Waal 2001).

Some behavioral patterns are common to both human and chimpanzee but not bonobo, as well as patterns common to both human and bonobo but not chimpanzee.

These behavioral patterns are either those of the LCA of *Pan* and *Homo*, which either chimpanzee or bonobo has lost, or patterns that were lacking in the LCA but evolved in both man and only one species of *Pan* in parallel after the split of chimpanzee and bonobo. (The last possibility is less likely from the viewpoint of parsimony.) In such cases, we tentatively show Category 1–3, the status of which remains ambiguous. The behavioral patterns of the LCA more likely are to be found in Categories (1), (1–2) or (1–3) than in others.

Preliminary comparisons have already suggested interesting contrasts between chimpanzee and bonobo (e.g., Mori 1983; Wrangham 1986; Nishida and Hiraiwa-Hasegawa 1987; Kano 1992; Stanford 1998b; Hohmann and Fruth 2003). If we scrutinize the minute details of the behavioral patterns of these sibling species, we may find many more differences between them.

As less is known of the behavior of the two chimpanzee subspecies of central Africa (e.g., Kuroda 1998; Sanz et al. 2004; Sanz and Morgan 2007), it is presently impossible to delineate Category (4) above. However, recent progress at Goualougo (Sanz et al. 2004; Morgan and Sanz 2006; Sanz and Morgan 2007, 2009; Sanz et al. 2009) is encouraging. Also, an increasing amount of information has emerged on West African chimpanzees, especially those of the Taï Forest (e.g., Boesch and Boesch-Achermann 2000), Bossou (Sugiyama 1998; Matsuzawa and Yamakoshi 1996; Sugiyama 2008), and Fongoli (Pruetz 2006; Pruetz and Bertolani 2007), in addition to older data from Kanka Sili (Kortlandt and Kooij 1963; Kortlandt and Bresser 1963; Albrecht and Dunnett 1971) and Assirik (McGrew et al. 1979; McGrew et al. 2003), we can tentatively regard the common patterns of eastern (Budongo, Gombe, Kanyawara, Mahale, Ngogo) and western (Bossou, Fongoli, Taï) chimpanzees as candidates for behavior common to the species *Pan troglodytes*. Since most chimpanzees kept in European zoos are of the west African subspecies, behavior of captive chimpanzees provide much information on this subspecies (e.g., de Waal 1982).

Behavioral patterns belonging to Categories 5, 6 or 7 are most likely to be cultural, but their status remains ambiguous, until it is determined by careful comparison whether the environmental factors that might explain the local differences are present or absent.

Methods

In compiling an audio-visual ethogram of Mahale chimpanzees, we mostly followed principles used by Nishida et al. (1999). First, we tried to make the ethogram maximally inclusive. Second, we distinguished functional definitions from 'anatomical' ones. Elements of the latter should be provided for all functionally-defined behavioral patterns. In order to attain this goal, we adopted the format of putting verbs in front of nouns or adjectives for the description of behavioral patterns. For example, we used "hunch bipedal" or "clip leaf" instead of "bipedal hunch" or "leaf-clip". Moreover, in this version, we used only a single 'noun' when it was more appropriate. For example, we listed "toy" instead of "play toy" because "toy" is enough to show that play is involved. Third, differently from the previous version, in this paper we included as many video clips as possible in the DVD. If we had no video clips for a particular behavior pattern, then we made effort to add illustrations of the behavioral pattern in the form of photographs, video-frames, or drawings, so that researchers at other study sites can verify whether the patterns that they see are the same or different from the ones described by us.

In order to classify behavioral patterns hierarchically and to describe social relationships, both 'anatomical' and functional terms are needed. This book's major aim is to list the names of all the behavioral elements in terms of anatomical details, but we also need functional terms in order to denote behavior that is meaningful in the natural environment. For example, <approach> is a functional term. Although it is simply <walk> or <run> from the anatomical perspective, when one chimpanzee comes closer to another, some interaction will likely occur between them. Therefore, we must have a term <approach> in addition to <walk>, and so on. Moreover, from the perspective of cognitive psychology, we need functional terms. For example, a chimpanzee who <detours> illustrates an important aspect of social cognition, in that the chimpanzee predicts from experience an associate's future behavior and changes her route accordingly, although <detour> may be nothing but <glance> plus <walk>. <Deceive> is a much more difficult functional term, as it entails an inference by the observer about the ape's intentions (Byrne and Whiten 1988). Deception may not be linked to any particular motor pattern (therefore, it is impossible or difficult to provide video clips or other illustrations), but if we ignore this term we will fail to capture an important part of the cognitive abilities of the chimpanzee.

T. Nishida et al., *Chimpanzee Behavior in the Wild: An Audio-Visual Encyclopedia*,
DOI 10.1007/978-4-431-53895-0_3, © Springer 2010

Some behavior occurs as a sequence of different behavioral patterns or behavioral complexes; the "charging display" is an example of such a sequence. Some terms designate several morphologically distinct behavioral patterns that occur in similar contexts and have similar functions, although they may not occur simultaneously. Thus, we divide behavior patterns into four 'types': Type A denotes simple, anatomical terms such as <walk bipedal>; Type B is for complex anatomical terms such as <wrestle>; Type C is for simple functional terms such as "approach"; and Type D is for complex functional terms such as "abuse". We designate the applicable Type for each term.

We have employed as many terms as possible from previous works, in particular Goodall (1968, 1986, 1989), van Hooff (1973), Bygott (1974), Berdecio and Nash (1981), Plooij (1984), Kano (1998) and Call and Tomasello (2007b), as well as Nishida et al. (1999). Many new terms, such as "groom self" are introduced here, but we do not seek to persuade readers to use neologisms. If we were to follow conventional usage and list terms by alphabetical order, then "groom" and "self-groom" would be placed many pages apart from each other although they designate the same motor pattern. To avoid such inconsistency, conventional terms were rearranged so that the 'action' was emphasized, and this policy of neologism was consistent with the 'verb first' principle outlined above.

Three-letter representations (codes) of behavioral patterns are useful in recording activities quickly and for displaying them economically in illustrations. Accordingly, we borrowed from Plooij (1984) when the appropriate abbreviation was available, but because Plooij's (1984) ethogram concerned mostly mother-infant behaviors, most of our abbreviations were new. Goodall's (1989) glossary descriptions were reproduced in the explanation of terms, if the behavioral patterns were the same as those seen at Mahale. The category numbers of (1)–(9) used in the glossary below follow the principles explained above in the Introduction.

Study Sites of Chimpanzees and Bonobos

Local names of the major study-sites referred in the Glossary are listed below with main reference sources. Shorter-term study sites such as Bai Hokou, Beni, Campo, Filabanga, Kabogo, Kanka-Sili, Kasakati, Moukalaba, Ndoki, Odzala, Petit Loango, and Tongo are described in the glossary.

Chimpanzees (*Pan troglodytes*)

P. t. verus

Assirik: Niokolo-Koba National Park, Senegal (McGrew et al. 1981)
Bossou: Guinea (Sugiyama 1981)
Fongoli: Senegal (Pruetz 2006)
Taï: Taï National Park, Cote d'Ivoire (Boesch and Boesch-Achermann 2000)
Tenkere: Sierra Leone (Alp 1997)

P. t. vellerosus

Ebo: Ebo Forest, Cameroon (Morgan and Abwe 2006)
Gashaka: Gashaka Gumti National Park, Nigeria (Fowler and Sommer 2007)

P. t. troglodytes

Goualougo : Nouabale-Ndoki National Park, Republic of Congo (Morgan and Sanz 2006)
Lopé: Lopé Reserve, Gabon (Tutin et al. 1991)

T. Nishida et al., *Chimpanzee Behavior in the Wild: An Audio-Visual Encyclopedia*,
DOI 10.1007/978-4-431-53895-0_4, © Springer 2010

P. t. schweinfurthii

Budongo: Budongo Forest Reserve, Uganda (Reynolds 2005)
Bwindi: Bwindi Impenetrable National Park, Uganda (Stanford and Nkurungi 2003)
Gombe: Gombe National Park, Tanzania (Goodall 1986)
Kahuzi: Kahuzi-Biega National Park, Democratic Republic of Congo (Yamagiwa et al. 1996)
Kalinzu: Kalinzu Forest Reserve, Uganda (Hashimoto and Furuichi 2006)
Kanyawara: Kibale National Park, Uganda (Wrangham and Peterson 1996)
Mahale: Mahale Mountains National Park, Tanzania (Nishida 1990)
Ngogo: Kibale National Park, Uganda (Mitani and Watts 1999)
Semliki: Toro-Semliki Wildlife Reserve, Uganda (Hunt and McGrew 2002)
Ugalla: Ugalla Forest Reserve, Tanzania (Itani 1979)

Bonobo (*Pan paniscus*) (all Democratic Republic of Congo)

Lilungu: (Bermejo et al. 1994)
Lomako: Lomako Forest Reserve (Susman 1984)
Lui Kotale: Salonga National Park (Hohmann et al. 2006)
Lukuru: (Thompson 2002)
Wamba: Luo Scientific Reserve (Kano 1992)
Yalosidi: (Kano 1983)

Catalogue

Classification of Behavior Patterns

1. Social interaction
2. Feeding and foraging
3. Expressive behavior
4. Positional behavior
5. Object manipulation
6. Culture
7. Others

1. Social interaction

    1-1. Agonism

        1-1-1. Aggression
        1-1-2. Threat and display
        1-1-3. Submission and defense
        1-1-4. Reassurance and appeasement

    1-2. Sexual behavior

        1-2-1. Copulation
        1-2-2. Courtship
        1-2-3. Others

    1-3. Maternal behavior

        1-3-1. Maternal/alloparental care and infant behavior
        1-3-2. Infant transport
        1-3-3. Weaning

    1-4. Grooming

        1-4-1. Social grooming and self-grooming
        1-4-2. Grooming posture and grooming request

T. Nishida et al., *Chimpanzee Behavior in the Wild: An Audio-Visual Encyclopedia*,
DOI 10.1007/978-4-431-53895-0_5, © Springer 2010

- 1-5. Social/solo play and play invitation
- 1-6. Inter-group relationships
- 1-7. Inter-species relationships
- 1-8. Other social interaction

2. Feeding and foraging
   - 2-1. Eat
   - 2-2. Techniques
   - 2-3. Food sharing
   - 2-4. Hunting

3. Expressive behavior
   - 3-1. Vocalization
   - 3-2. Facial expression
   - 3-3. Gesture

4. Positional behavior
   - 4-1. Locomotion
   - 4-2. Posture

5. Object manipulation
   - 5-1. Hold and transport object
   - 5-2. Tool-use to feed
   - 5-3. Tool-use in threat and aggression
   - 5-4. Tool-use in other contexts

6. Culture

7. Others
   - 7-1. Physiological behavior
   - 7-2. Self-directed behavior
   - 7-3. Movement and grouping
   - 7-4. Inspection
   - 7-5. Maintenance

1-1-1. Agonism: Aggression

[abuse] [abuse carcass] [aggress] [aggress in sexual frustration] [aggress morally] [aggress, redirected] [attack] [attack concertedly] [attack human] [battle] [bite] [brush aside] [chase] [club] [club another] [compress lips] [drag by hand] [drag other by hand] [emasculate] [fight] [grab] [grapple] [hit] [immobilize] [interact with human] [kick] [kick other] [kill] [kill adolescent male] [kill adult female] [kill adult male] [kill infant] [kill juvenile] [leap on] [ostracize] [pinch aggressively] [press] [press down] [push] [push away] [retaliate]

[roll] [scratch aggressively] [show off] [slap other] [spar] [stamp other] [support] [support dominant] [support older] [support subordinate] [swing and kick]

1-1-2. Agonism: Threat and display

[aggress] [bark] [bark, infantile] [beat chest] [break branch] [break tree] [charge] [clap hand] [compress lips] [cough bark] [display as contest] [display away] [display past] [display toward] [display, charging] [display, rain] [display, streambed] [drag by hand] [drag carcass by hand] [drag dry leaves] [drum] [flail] [flail arm] [gaze] [harass] [hit bush bipedal] [hit toward] [hoot] [hunch] [hunch and sit] [hunch bipedal] [hunch over] [ignore] [intervene] [intervene to separate] [kick backward quadrupedal] [kick bipedal] [kick buttress] [knock down with both arms] [knock down with one arm] [leap bipedal with squared shoulders] [lift and drop] [police] [protest] [pull rock to roll] [push backward] [push down] [push forward] [raise arm quickly] [rock back and forth] [rock side to side] [scramble for food] [scratch dry leaves] [shake arm, abduct] [shake back and forth with four limbs] [shake branch] [shake branch up and down with feet] [shake long object irregularly] [shake object up and down] [show off] [slap buttress or tree trunk] [slap ground] [slap self] [slap-stamp] [slap wall] [snub] [stamp] [stamp bipedal] [stamp quadrupedal] [stamp trot] [stare fixedly] [strip leaf] [supplant] [swagger bipedal] [sway woody vegetation] [swing] [swing and kick] [tease] [threaten] [throw] [throw at] [throw at animate object] [throw dry leaves] [throw sand] [tip head] [waa bark]
(See also 5-3. Tool-use in threat and aggression)

1-1-3. Agonism: Submission and defense

[appease] [bend away] [bob] [bob bipedal] [bow] [crouch] [dab] [escape] [flee] [glance] [greet] [grin] [hesitate] [hop bipedal on spot] [hop quadrupedal on spot] [pant-grunt] [pant-grunt with bent elbow] [parry] [present with limbs extended] [present with limbs flexed] [scream] [shake hand side to side quickly] [shake other] [struggle free] [throw dry leaves] [vacate]

1-1-4. Agonism: Reassurance and appeasement

[appease] [bend and stretch knee] [confiscate] [console] [embrace full] [embrace half] [extend hand] [grasp hand] [greet] [hold genitals] [hug] [kiss] [kiss with open mouth] [kiss with pout face] [kiss with tongue] [massage shoulder] [mount] [mount, misdirected] [mouth] [nod and mouth] [offer arm] [pat] [poke] [press teeth against back] [push finger into mouth] [put heel on back] [put rump to rump] [reach wrist toward] [reassure] [reconcile] [shake penis] [solicit reassurance contact] [solicit support] [suck in reassurance] [take finger in mouth] [thrust] [thrust bipedal] [thrust, misdirected] [touch] [touch scrotum]

1-2-1. Sexual behavior: Copulation

[copulate] [copulate between group] [copulate dorso-ventral] [copulate ventro-ventral] [dart] [incest] [interfere in copulation] [mount in copulation] [pant in copulation] [squeal in copulation] [thrust]

1-2-2. Sexual behavior: Courtship

[bend shrub] [bend shrub in courtship] [clip leaf by hand] [clip leaf by mouth] [club ground] [dance orthograde] [hop bipedal on spot] [hop quadrupedal on spot] [hunch] [hunch and sit] [hunch bipedal] [lift and drop] [open thighs] [present with limbs extended] [present with limbs flexed] [pull through in courtship] [push] [raise arm quickly] [rap] [scratch dry leaves] [shake branch] [shake detached branch] [slap branch] [solicit copulation] [stamp bipedal] [stand bipedal] [strip leaf] [swagger bipedal] [swagger on knuckles] [sway woody vegetation] [tap heel] [throw branch] [thrust bipedal] [touch]

1-2-3. Sexual behavior: Others

[adduct penis] [adolescent infertility] [adolescent swelling] [aggress in sexual frustration] [avoid incest] [check penile erection] [consort] [eat semen] [ejaculate] [erect penis] [estrus] [flabby bottom] [fumble clitoris] [fumble penis] [fumble penis with foot] [herd] [herd in coalition] [inspect genitals] [lead] [masturbate] [rub genitals] [rub genitals to substrate] [semen] [sexual skin] [shake penis] [swelling of sexual skin] [thrust in vacuum]

1-3-1. Maternal behavior: Maternal/alloparental care and infant behavior

[abandon] [adopt] [aid in locomotion] [balance] [bounce foot/feet up and down rhythmically] [care alloparentally] [care alloparentally for another species] [care maternally] [change nipple] [clamp elbow] [cling] [cradle with hand] [cradle with leg] [extend arm as ladder] [extend leg as ladder] [gestation period] [give birth] [grasp] [hug] [kidnap] [kiss with pout face] [lie and hug] [mother-offspring relations] [mummify] [nuzzle] [orphan] [pat] [pregnancy] [protect] [rescue] [respond to dead chimpanzee] [retrieve infant] [return infant] [rush to embrace] [scan] [share bed] [share food] [spring] [suck] [suck in reassurance] [suckle] [travel alone after childbirth] [wedge]

1-3-2. Maternal behavior: Infant transport

[crutch] [grasp and push shoulders] [put dorsal] [put dorsal from ventral] [put ventral] [put ventral from dorsal] [ride bipedal] [ride clinging] [ride dangling] [ride dangling and touch dry leaves] [ride dorsal] [ride jockey] [ride prone] [ride quadrupedal] [ride supine] [ride ventral] [ride ventral with limb extended] [scoop infant] [solicit riding] [step up on leg] [touch] [transport corpse of infant] [transport two offspring] [transport with hand support] [transport with thigh support]

1-3-3. Maternal behavior: Weaning

[approach and withdraw] [brush aside] [contact with nipple] [cover nipple] [distract] [drop infant] [fall over backward] [fend] [ignore] [interfere in copulation] [leave to protest] [lower head and shoulder] [lower rump] [monitor mother] [push ahead] [push away] [reject] [reject infant] [reject-move] [reject-sit] [scatter dry leaves] [shake rump] [shrug] [stand bipedal] [stare fixedly with head down and bottom up] [throw temper tantrum] [turn around] [wean] [whimper] [whimper-scream]

1-4-1. Grooming: Social grooming and self-grooming

[allow] [clack teeth] [crush] [gaze] [groom] [groom briefly] [groom by hand] [groom carcass] [groom leaf] [groom mutually] [groom object or substrate] [groom reciprocally] [groom self] [groom unilaterally] [groom with mouth] [groom wound] [groom, dyadic] [groom, polyadic] [groom-branch-clasp] [groom-hand-clasp] [groom-hand-clasp unilaterally] [inspect leaf] [presbyopia] [remove] [remove lice] [remove sand flea] [scrape] [scratch] [scratch socially] [scratch socially, poke type] [scratch socially, stroke type] [smack lip] [sputter] [squash ectoparasite on arm] [squash leaf]

1-4-2. Grooming: Grooming posture and grooming request

[grab] [groom-branch-clasp] [groom-hand-clasp] [groom-hand-clasp unilaterally] [hold head or face] [lean forward] [lie supine with legs apart] [lie with back to another] [look back] [lower arm] [lower head] [lower leg] [offer arm] [offer back] [pull] [pull face to face] [pull head with hands] [push] [raise and hold leg] [raise arm slowly] [raise arm to hold branch] [raise arm with elbow bent] [raise leg while lying] [raise other's arm] [raise other's chin] [raise other's leg] [scratch dry leaves] [scratch self signalling] [sit and turn back] [sit behind individual] [sit face to face] [sit sideways] [solicit grooming] [solicit grooming turn] [stare fixedly with head down and bottom up] [touch] [turn face upward] [turn up lip]

1-5. Social/solo play and play invitation

[backstroke] [balance] [beat sole with palm] [bend shrub] [bite and pull hairs] [bounce foot/feet up and down rhythmically] [chase] [circle orthograde] [circle orthograde off ground] [circle quadrupedal] [club other] [collect dry leaves] [contact water][cover self] [drag and circle] [drag by hand] [drag by mouth] [drag dry leaves] [drag other by hand] [drag other by mouth] [drape] [drip] [drop] [drop bark bits] [drop self] [dunk face] [enter into hole] [fend] [flail] [flail arm] [flee] [flop] [glove] [grab] [grab and shake] [grapple] [grasp and heave] [grasp hand] [handicap self] [hang and spin] [hang and stamp] [hang in sloth position] [hang upside-down by feet] [hang upside-down by hands] [hang with legs pitterpat] [hang-wrestle] [headstand] [hit and run] [hit bush bipedal] [hit ground with fist] [hold body part in mouth] [hop quadrupedal on spot] [hurl self] [interact with human] [interfere play] [join play] [kick] [kick back] [kick heel] [kick other] [kill time] [kick up] [leap down and wait] [leap on] [leap up] [lie and watch] [lie on other] [lie supine] [lie supine and shake arms and legs] [lift and drop] [look at water] [look between thighs] [mount, misdirected] [mouth] [mouth and shake] [nod and mouth] [nod to water surface] [nod to water surface and mouth water] [nod with body part in mouth] [nod with object in mouth] [nod with play face] [pass under] [pick up and release] [pile dry leaves] [pinch clitoris] [pirouette] [play] [play bite] [play face] [play face, full] [play face, half] [play, imaginary] [play in bed] [play-pant] [play, parallel] [play, rough and tumble] [play socially] [play socially with object] [play solo] [play walk] [play with another animal] [play with object] [play with sand] [play with urine] [play with water] [poke] [poke with play face] [press] [press down] [press neck with lower arm] [press object on] [pull] [pull down] [pull each other] [pull leaf-pile] [pull object from opposing sides] [pull rock to roll] [pull with mouth] [push] [push forward] [push head into ventral] [push leaf-pile] [push/ pull swing] [put face to] [put mouth into water] [rake] [rebuff play] [release infant to fall] [ride dangling and touch dry leaves] [roll] [rotate fruit] [rub dorsum to conspecific] [rub object to body] [scatter dry leaves] [scratch dry leaves] [shake branch up and down with feet] [shake face side to side] [shake face with object in mouth] [shake head] [shake object up and down] [shake off] [shake other] [shake rock up and down with feet] [sit on] [slap in invitation] [slap other] [slap self] [slide down boulder] [solicit play] [solicit play with object] [solicit play with object in mouth] [somersault] [somersault, backward] [somersault, backward with dry leaves] [somersault, forward] [somersault, side] [spar] [spit water] [splash water] [stamp] [stamp in invitation] [stamp other] [stamp quadrupedal] [stamp water] [stand with head down, bottom up] [step on] [stir water] [struggle free] [suspend and shake up and down] [swing] [swing above] [swing and kick] [swing forward and upward] [tag] [tease] [throw] [throw at] [throw at animate object] [throw dry leaves] [throw sand] [throw stone or rock] [thrust] [thrust in vacuum] [thrust, misdirected] [tickle] [tickle self with object] [tilt head] [toy] [trample] [travel and play] [trifle with] [walk in sloth position] [walk quadrupedal on backs of hands] [wrestle] [wrestle bipedal] [wrestle with fingers]

1-6. Inter-group relationships

[attack concertedly] [battle] [copulate between group] [emigrate] [follow specific female] [immigrate] [inter-group behavior] [intrude] [kill adult male] [kill infant] [patrol] [respond to neighboring unit group] [stalk] [territory] [transfer] [unit group] [visit]

1-7. Inter-species relationships

[abuse carcass] [attack human] [bark] [care alloparentally for another species] [catch with hand] [clap hand] [club ground] [drop branch] [flee from colobus male] [groom carcass] [habituated] [hide] [interact with human] [kill another species] [play with another animal] [respond to baboon] [respond to dead animal] [respond to leopard] [respond to lion] [respond to predator] [throw at] [throw at animate object] [trifle with] [whisk fly with arm] [whisk fly with leafy stick] [wraa]
(See also 2-4. Hunting)

1-8. Other social interaction

[alliance] [coalition] [conflict behavior] [deceive] [friendship] [imitate] [interfere] [interfere copulation] [interfere fishing] [interfere play] [misunderstand] [reject] [restrain] [snatch] [steal] [stroke] [teach] [turn face away] [wait turn]

2-1. Feeding and foraging: Eat

[chew] [disperse seed] [drink] [drink from hole in tree] [drink from lake] [drink from stream] [eat] [eat algae] [eat ant] [eat beetle larva] [eat blossom] [eat *Camponotus* ant] [eat carcass] [eat *Crematogaster* ant] [eat *Dorylus* ant] [eat egg] [eat eye mucus] [eat feces] [eat fruit, inner skin] [eat fruit, pulp] [eat gall] [eat honey] [eat

infant] [eat insect] [eat leaf] [eat meat] [eat nasal mucus] [eat *Oecophylla* ant] [eat petiole] [eat phloem] [eat pith] [eat resin] [eat rock] [eat root] [eat seed] [eat semen] [eat termite] [eat termite soil] [eat xylem] [grunt, food] [lick] [lick lips] [lick rock] [lick wood] [lick wound] [reingest vomit] [scream, food] [suck]

2-2. Feeding and foraging: Techniques

[break bone at joint] [break branch] [break branch with foot] [break open by pushing chin] [break tree] [brush away from branch] [brush away from self] [catch with hands] [dig] [dig for underground storage organ by hand] [dig for water by hand] [dip fruit wadge into water] [dip hand and lick water] [discard fruit skin with mouth] [dunk] [eat with foot] [fill mouth with food] [grope] [handedness] [keep water in mouth] [mop ant] [peel with hand] [peel with teeth] [pick out pulp] [pick up discarded food] [process food] [pull down] [pull out] [pull through with hand] [pull through with mouth] [remove objects from water surface] [rinse] [scrape] [scrub pelt] [shake wet arm to catch termites] [sniff fruit] [spit juice] [spit seed] [squeeze] [stand bipedal] [store] [swallow leaf] [swallow seed] [tear] [throw fruit against rock] [touch fruit] [transport food] [twist] [wadge] [wadge by adding leaf] [wadge without adding leaf] [wash] [wash colobus skin]
(See also 5-2. Tool-use for feeding)

2-3. Feeding and foraging: Food sharing

[allow] [beg] [clamp elbow] [extend hand to beg] [fend] [give] [mouth for begging] [peer] [push away] [share food] [take] [turn around] [twist]

2-4. Feeding and foraging: Hunting

[chase] [drag to kill] [dunk] [eat meat] [expel] [hunt] [hunt with tool] [knock with both arms] [knock with one arm] [monitor monkeys] [scrub pelt] [share food] [stalk] [twist] [wash] [wash colobus skin]

3-1. Expressive behavior: Vocalization

[bark] [bark, infantile] [choke in tantrum] [clack teeth] [cough bark] [distress call] [grunt] [grunt, aha] [grunt, extended] [grunt, food] [grunt in bed] [hoo] [hoot] [huu] [mass excitement] [pant] [pant-bark] [pant-grunt] [pant-hoot] [pant in copulation] [play-pant] [scream] [scream, food] [smack lip] [sputter] [squeal in copulation] [staccato call] [vocalize] [waa bark] [whimper] [whimper-scream] [wraa]

3-2. Expressive behavior: Facial expression

[compress lips] [flip lip] [funny face] [grin] [grin-full-closed] [grin-full-open] [grin-low-closed] [grin-low-open] [hoot face] [play face] [play face, full] [play face, half] [pout] [protrude tongue] [pucker cheek] [relaxed face] [sneer]

3-3. Expressive behavior: Gesture

[beat chest] [bend and release] [bend and stretch knee] [bend away] [bob] [bob bipedal] [bow] [clap hand] [clip leaf by hand] [clip leaf by mouth] [dab] [extend] [extend hand] [extend hand and put knuckle on ground] [extend hand to beg] [extend hand to hold tree] [extend hand, palm downward] [extend hand, palm sideways] [extend hand, palm upward] [extend leg] [fall over backward] [flap] [hit ground with fist] [hit toward] [lower arm] [lower head] [lower leg] [make sound] [offer arm] [open thighs] [pull through in courtship] [raise and hold leg] [raise arm(s) bipedal] [raise arm quickly] [raise arm slowly] [raise arm to hold branch] [raise arm with elbow bent] [raise leg while lying] [rap] [reach wrist toward] [scratch dry leaves] [scratch self signalling] [shake arm, abduct] [shake arm, adduct] [shake back and forth with four limbs] [shake branch] [shake branch up and down with feet] [shake detached branch] [slap] [slap branch] [slap buttress or tree trunk] [slap ground] [slap in invitation] [slap self] [slap-stamp] [slap wall] [solicit companion] [solicit copulation] [solicit grooming] [solicit play] [solicit reassurance contact] [solicit riding] [solicit support] [stamp] [stamp bipedal] [stamp in invitation] [stamp quadrupedal] [strip leaf] [throw temper tantrum] [tip head]

4-1. Positional behavior: Locomotion

[backstroke] [brachiate] [climb] [climb cliff] [climb vertical] [climb vertical, extended elbow] [crutch] [descend] [descend by brachiating] [descend orthograde feet first] [descend tree trunk feet first] [descend tree trunk head first] [drop self] [fall] [gallop] [hop bipedal on spot] [hop quadrupedal on spot] [leap] [leap between trees] [leap between trees with object] [leap bipedal] [leap bipedal with squared shoulders] [leap down] [leap quadrupedal] [leap up] [limp] [locomote] [pirouette] [play walk] [retreat] [retreat bipedal] [run] [run bipedal] [run quadrupedal] [run tripedal] [slide down vertically] [somersault] [somersault, backward] [somersault, forward] [somersault, side] [stamp trot] [stumble] [sway and move] [swing] [swing and grasp] [swing forward and upward] [tumble] [wade] [walk bipedal] [walk in sloth position] [walk quadrupedal on backs of hands] [walk quadrupedal on knuckles] [walk quadrupedal on palms] [walk tripedal]

4-2. Positional behavior: Posture

[cling] [cross] [cross arms] [cross arms on head] [cross legs] [crouch] [dangle] [groom-branch-clasp] [groom-hand-clasp] [hang] [hang in sloth position] [hang tripedal] [hang upside-down by feet] [hang upside-down by hands] [hang-stand] [headstand] [hunch] [hunch and sit] [hunch bipedal] [hunch quadrupedal] [lean] [lean forward] [lie] [lie and hug] [lie lateral] [lie prone] [lie-sit] [lie supine] [lie supine with legs apart] [lie with legs crossed] [raise] [raise and hold leg] [raise arm to hold branch] [raise leg while lying] [sit] [sit and lean] [sit face to face] [sit prone] [squat] [stand bipedal] [stand quadrupedal] [stand quadrupedal heel up] [stand with head down, bottom up]

5-1. Object manipulation: Hold and transport object

[discard] [drag by hand] [drag carcass by hand] [drag carcass by mouth] [grasp] [hold object in groin pocket] [hold object in mouth] [hold object in neck pocket] [hold object on head] [leap between trees with object] [lift] [lift and drop] [lift rock] [move] [pick up] [suspend] [transport] [transport bipedal] [transport corpse of infant] [transport food] [transport in foot] [transport in groin pocket] [transport in

hand] [transport in mouth] [transport in neck pocket] [transport on back] [transport on head or nape] [transport on shoulder] [transport quadrupedal]

5-2. Object manipulation: Tool-use to feed

[brush-tip fishing probe] [clip leaf for fishing probe] [dig for army ant with hand/stick] [dig for underground storage organ with tools] [dig for water with stick] [dig stingless bees' nest with stick] [dig termite nest with stick] [dip ant and wipe] [dip ant directly] [dip fluid] [expel] [extract] [fish] [fish for carpenter ant] [fish for termite] [fold leaf] [hammer nut with stone] [hammer nut with wood] [hunt with tool] [leaf-midrib spoon] [leaf-sponge] [leaf-spoon] [lever open] [make tool] [mop ant with leaves] [perforate] [pick out bone marrow] [pick out nutmeat] [pound] [pound food on object] [pound food on wood] [pound insect] [pound pestle] [pound resin] [prop anvil] [puncture] [push and pull sponge] [push object into] [scoop algae] [tool] [use tool]

5-3. Object manipulation: Tool-use in threat and aggression

[club] [club ground] [club other] [drag branch] [drop] [drop branch] [flail long object] [throw at inanimate object] [throw branch] [throw splash] [throw stone or rock] [tool] [use tool]

5-4. Object manipulation: Tool-use in other contexts

[bed] [bend branch] [clip leaf by hand] [clip leaf by mouth] [comb] [cover from rain] [dab leaf] [drape] [groom leaf] [hook branch] [make bed] [make cushion] [make day bed] [make ground bed or cushion] [make leaf cushion] [make night bed] [make tool] [play socially with object] [play with object] [probe] [probe nasal passage] [push object into] [push to shoulder] [seat-stick] [sniff with tool] [solicit play with object] [solicit play with object in mouth] [squash leaf] [stepping-stick] [tickle self with object] [tool] [tool composite] [tool set] [toy] [use tool] [whisk fly with leafy twig] [wipe with detached object]

6. Culture

[bend and release] [bend shrub in courtship] [clip leaf by hand] [clip leaf by mouth] [comb] [cover from rain] [culture] [dab leaf] [dig for army ant with hand/stick] [dig for underground storage organ by hand] [dig for underground storage organ with tools] [dig stingless bees' nest with stick] [dig termite nest with stick] [dip ant and wipe] [dip ant directly] [dip fruit wadge into water] [eat algae] [eat ant] [fashion] [fish for carpenter ant] [fish for termite] [fold leaf] [groom leaf][groom-hand-clasp] [hammer nut with stone] [hammer nut with wood] [inspect leaf] [leaf-sponge] [lever open] [lick rock] [lick wood] [make leaf cushion] [perforate] [pick out bone marrow] [pick out nutmeat] [pound food on object] [pound food on wood] [pound insect] [pound pestle] [pound resin] [prop anvil] [pull leaf-pile] [puncture] [rub muzzle] [scoop algae] [scratch socially, poke type] [scratch socially, stroke type] [sputter] [squash ectoparasite on arm] [squash ectoparasite on palm] [squash leaf] [strip leaf] [tap heel] [throw fruit against rock] [throw splash] [tradition] [whisk fly with leafy stick] [wipe with detached object]

7-1. Physiological behavior

[adduct penis] [bristle] [close eyes] [cough] [death] [defecate] [defecate, prone] [defecate quadrupedal] [defecate, sitting] [diarrhea] [die] [ejaculate] [erect penis] [fart] [hiccup] [open eyes] [parry] [senescence] [shiver] [sleep] [sneeze] [sniffle] [urinate] [urinate, prone] [urinate quadrupedal] [urinate, sitting] [vomit] [wriggle] [yawn]

7-2. Self-directed behavior

[bite self] [clasp self] [fumble clitoris] [fumble nipple] [fumble penis] [fumble penis with foot] [inspect self] [lick wound] [masturbate] [medicate self] [pick ear] [pick nose] [probe nasal passage] [push peri-anogenital region with finger] [scratch self] [slap self] [suck self] [suck thumb] [suck toe] [tickle self with object]

7-3. Movement and grouping

[approach] [approach and withdraw] [arise] [arrive] [avoid] [avoid fallen log] [bend away] [chase] [core area] [creep] [dart] [depart] [depart together] [detain] [detour] [family] [fission and fusion] [flee] [flee after startle] [follow] [follow in contact] [follow specific female] [go ahead] [group] [herd] [hesitate] [hide] [lead] [leap up in surprise] [leap into lap] [leave] [leave and climb] [party] [pass] [pass under] [pass with body contact] [regulate direction] [rest] [reunite] [reverse] [scan] [search for conspecific] [separate] [solicit companion] [stalk] [stand quadrupedal heel up] [step over] [supplant] [territory] [travel] [travel alone after childbirth] [vacate] [wade] [wait for companion]

7-4. Inspection

[crowd] [gaze] [glance] [grope] [inspect] [inspect fruit] [inspect genitals] [inspect leaf] [inspect self] [inspect wound] [listen] [look around] [look at water] [look back] [look between thighs] [look up] [monitor] [monitor monkeys] [monitor mother] [patrol] [peep] [peer] [peer together] [probe] [probe with finger] [respond to dead animal] [respond to dead chimpanzee] [rummage] [scan] [search for conspecific] [search for object] [sniff] [sniff finger] [sniff fruit] [sniff mouth] [sniff with tool] [stand bipedal] [stare fixedly] [tilt head] [touch fruit] [touch with foot] [vigilance] [watch]

7-5. Maintenance

[brush away from self] [catch with hand] [comb] [cover from rain] [dab leaf] [eat eye mucus] [lick wound] [medicate self] [pick ear] [pick nose] [probe nasal passage] [push peri-anogenital region with finger] [rub dorsum] [rub hand or foot] [rub hand with hand] [rub muzzle] [scratch self] [shake head] [shelter from rain] [sickness] [whisk fly with arm] [whisk fly with leafy stick] [wipe with detached object]

Remarks

1. Principle of text descriptions in the catalogue goes as follows:

(1) Start with first-person singular verb whenever possible
(2) Assume 'chimpanzee' as subject of verb
(3) Omit articles ('a', 'an', 'the') for conciseness
(4) Omit verb "to be"
(5) Eliminate redundancy, e.g. 'one hand' to 'hand'

2. Use of symbols, etc.

(1) Behavioral patterns listed in this book are bracketed by < >
(2) Local behavioral patterns that have been recorded at other sites, but not at Mahale, are shown in bold italics and are not categorized
(3) In addition to the behavioral catalogue, we include some basic terms (nouns and adjectives) such as "Infant" and place names of study sites, so readers can use this book as a dictionary. Such terms are discriminated from behavioral patterns by using bold font
(4) Synonyms are in italics

T. Nishida et al., *Chimpanzee Behavior in the Wild: An Audio-Visual Encyclopedia*,
DOI 10.1007/978-4-431-53895-0_6, © Springer 2010

Glossary

A

abandon (ABD)

Nishida (1983a): Alloparent leaves infant charge and departs without returning it to its mother. (So, infant must return to its mother by itself or be retrieved.) Also occurs in bonobos (Kano unpublished). (Type D) Category 2. See Fig. 1.

Fig. 1 Abandon: Five-year-old female, Blondy (*left*), left infant charge (*right, behind*) 10 m away from its mother (*right*) (T. Nishida)

abuse (ABU)

Nishida (1983a): Older animal apparently mistreats an infant by <brush aside>, <push away>, <roll>, <kick>, <push>, <stamp>, <hit>, <pull>, <pull> foot, <press> legs, etc. Mother may abuse her infant. Also seen in bonobos of Wamba (Kano unpublished). (Type D) Category 1. See video.

abuse carcass (ABC)

Behavior directed to dead animal without eating it after chimpanzees have killed it. Apparently not directly related to consumption of meat.

T. Nishida et al., *Chimpanzee Behavior in the Wild: An Audio-Visual Encyclopedia*,
DOI 10.1007/978-4-431-53895-0_7,

Include <slap>, <stamp>, <kick>, <knock with both arms>, <knock with one arm>, <throw>, <throw branch>, <bite>, <suspend>, <shake object up and down>, and <sit on>. Colobus and leopard carcasses were abused at Mahale. However, chimpanzees also <hug>, <kiss with pout face>, <groom>, <lie on other>, and even <share day bed>. Cf. <respond to dead animal>, <trifle with>, and <care alloparentally for another species>. (Type D) Category 3–7. See video.

adduct penis (ADP)

Goodall (1989): "Erect penis spasmodically jerks" up and down. See also <erect penis>. (Type A) Category 1. See video.

adolescent

Sexually mature but socially immature (Hayaki 1985a, b; Pusey 1990). Males 9–15 years old, Females 9–13 years old. See Fig. 2.

Fig. 2 Adolescent: Ten-year-old male of K group, Masisa (T. Nishida)

adolescent infertility

Adolescent females usually do not conceive although they copulate often.

adolescent sterility

See <adolescent infertility>.

adolescent swelling

Goodall (1989) Adolescent swelling: "The first tiny swellings of the sex skin initially involve only the labia. These gradually become larger and more regular over the next 3 years or so until the first adult-sized swelling attracts the attention of the adult males…" See Fig. 3.

Fig. 3 Adolescent swelling: Nine-year-old female, Abi (T. Nishida)

adopt (ADO)

Long-continuing alloparental care of orphaned infant by single alloparent, sometimes lasting more than a year. At Mahale, at least five older adolescent or young adult nulliparous females seen to <hug>, <groom>, <transport>, <protect>, <play>, <share food>, and <share bed> with orphan under 5 years old. Long-term (>1 year) adoption by adult males was reported from Taï (Boesch 2009). In bonobos, care of 2-year-old infant by newly immigrated female seen for one week (Kano 1998). (Type D) Category 1. See Fig. 4.

Fig. 4 Adopt: Ten-year-old female, Tula, adopted 3-year-old female, Maggie, and cared for her for more than 2 years (T. Nishida)

adult

Sexually and socially mature. Males ≥16 years old, Females ≥14 years old. See Fig. 5.

Fig. 5 Adult: Eighteen-year-old male, Nsaba (T. Nishida)

aeroplane

See <balance>.

aggress (AGG)

<attack> or <threaten>. (Type D) Category 1. See video. See also video "aggress, redirected".

aggress in redirection

See <aggress, redirected>.

aggress in sexual frustration (AGF)

Aggress to sexual partner in response to refusal to copulate. (Type D) Category 1–5. See video.

aggress morally (AGM)

de Waal (1991) Moralistic aggression. Chimpanzee retaliates against companion after 'betrayal'. When alpha male supported adult female against his subordinate male ally during competition for meat, latter chased alpha male (Nishida 1994). (Type D) Category 1.

aggress, redirected (ARE)

Aggress to another (usually subordinate) individual after being aggressed against by more dominant individual. Goodall's (1989) Redirection of aggression. Kano's (1998) Redirection. (Type D) Category 1. See video.

aggressive scratch

See <scratch aggressively>.

aha-grunt

See <grunt, aha>.

aid in combat
See <support>.

aid in locomotion (AIL)
Locomotory aid: Mother or other individual assists its charge by extending arm or leg and retrieving infant stranded in tree, e.g., if it cannot cross gap. Cf. <extend arm as ladder>: allomother who is below infant's mother extends arm upwards to mother to facilitate infant's return to her. (Type D) Category 1. See videos. See also video "fall".

aimed throw
See <throw at>.

algae-scoop
See <scoop algae>.

alliance
Short-term and expedient mutual support against common rival. Cf. <coalition>. See video.

alloparental care
See <care alloparentally>.

allow (ALW)
Individual A permits individual B to approach, make contact with A and perform act of concern to A. For example, A allows B to suck A's wound, to groom A, to take food from A's control, to kidnap A's infant, etc. May include <sit>, <stand quadrupedal>, <watch>, etc. (Type D) Category 1. See video.

alpha male
Most dominant male of unit group or community. All other individuals pant-grunt to him. See video.

ant-dip-single
See <dip ants directly>.

ant-dip-wipe
See <dip ants and wipe>.

ant-fish
See <fish for carpenter ants>.

anvil-prop
See <prop anvil>.

appease (APS)
Goodall (1989): "Make submissive gestures directed toward a dominant individual after aggression, or in an attempt to prevent aggression." Includes <pant-grunt>, <touch>, <extend hand>, <kiss>, <embrace half>, <present with limbs extended>, <present with limbs flexed>, <groom briefly>, etc. Bonobos at Wamba appease by <touch>, <embrace

half>, <present>, <groom briefly>, etc. (Kano unpublished). (Type D) Category 1.

approach (APP)
Goodall (1989): "Individual moves toward another." Plooij's (1984) Approach (APP). van Hooff (1973) divided approach into "smooth" and "hesitant." Includes <walk quadrupedal>, <walk bipedal>, and <run>. Cf. <leave>. (Type C) Category 1. See video.

approach and withdraw (APW)
Infant, especially weanling, repeatedly approaches and withdraws from mother often with <whimper> when wanting to suck from mother. Estrous female may also do so when soliciting mating from male. Cf. <leave to protest> and <monitor mother>. (Type B) Category 3–5. See video.

arise (ARS)
Shift from reclining posture to sitting or standing. Rising by groomee stimulates groomer to resume grooming. (Type A) Category 1–2. See video.

arm ladder
See <extend arm as ladder>.

arm round
See <embrace half>.

arm stretch
See <extend hand, palm downward>.

arm swing
See <brachiate>.

arm threat
See <raise arm quickly>, <shake arm> and <hit toward>.

arrive (ARV)
Goodall (1989): "Chimp arrives and joins the group." Often accompanied by greeting. (Type C) Category 1. See video.

Assirik
Study area of *Pan troglodytes verus* in Niokolo-Koba National Park, Senegal (12°50′N, 12°45′W). Research done by W.C. McGrew and C.E.G. Tutin from 1976 to 1979 (McGrew et al. 1988).

attack (ATT)
Goodall (1989) Attack: "Aggressive physical contact between two, or more, individuals." Includes <charge>, <push>, <hit>, <grab>, <scratch>, <pinch>, <slap>, <drag>, <kick>, <stamp>, and <bite>. Cf. <battle>. Also in bonobos (Kano 1998). (Type D) Category 1. See video.

attack concertedly (ATC)

Nishida et al. (1995) Gang attack. Two or more chimpanzees simultaneously attack one or two others. Occurs both within (e.g. Takahata 1985; Nishida and Hiraiwa-Hasegawa 1985; Nishida 1994; Nishida et al. 1995; Goodall 1986, 1992) and between groups (Goodall 1986; Nishida 1979; Boesch and Boesch-Achermann 2000). See also <ostracize>. Also in bonobos of Wamba (Kano unpublished). (Type D) Category 1. See video.

attack human (ATH)

Wild chimpanzees occasionally attack human infants and juveniles and kill and eat them. Recorded from Gombe (Goodall 1986; Kamenya 2002) and near Budongo (Reynolds 2005). Habituated adult male chimpanzees occasionally push, slap or kick human observers as part of charging displays (Kamenya 2002; Nishida et al. 2009). Non-habituated adult female once bit two adult male humans (researchers) fiercely when her infant was being attacked by M group males at Mahale (Kutsukake and Matsusaka 2002). See also <interact with human>, <play with another animal> and <eat meat>. (Type D) Category 3–5.

auto-groom

See <groom self>.

avoid (AVD)

Locomote away from another. Plooij's (1984) Avoid (AVO), Goodall's (1989) Avoidance, van Hooff's (1973) Avoidance. See <detour>, <creep> and <hide>. Kano's (1998) Avoidance for bonobos. (Type D) Category 1. See video.

avoid fallen log (AVL)

When fallen log blocks path, infant moves onto log, then resumes ride on her back after mother passes under log. (Type B) Category 2–3. See video.

avoid incest (AVI)

Does not mate with close relatives such as parent, sibling or offspring. Incest avoidance known from many mammals and birds. See <incest>. (Type D) Category 1.

B

baboon

See <respond to baboon>.

backstroke (BST)

Solo locomotor play. Infant or young juvenile slides supine downwards, on slope covered with dry leaves, head first while kicking ground with feet, while keeping both arms ahead. Performed by juvenile male, Michio, and infant female, Athena. (Type A) Category 8–9. See video.

Bai Hokou

Study area of *Pan troglodytes troglodytes* in Dzanga-Sangha region of Central African Republic (2°50′N, 17°20′E). Research done by RW Carroll in 1988–1989 (Fay and Carroll 1994).

balance (BAL)

Nishida (1983a) Aeroplane, Plooij's (1984) Balance (BAL): Mother or adult/adolescent alloparent lying supine holds (usually prone) infant and lifts infant above ground by one to four of its limbs. Caretaker may gently bounce infant up and down with foot or feet. Caretaker may grasp one or both of infant's hands. Similar behavior by bonobo (Kano 1998; de Waal 1995, p 64) and gorilla (Whiten 1999) mothers. (Type B) Category 1. See video.

bang

See <throw fruit against rock>.

bare teeth

See <grin>.

bark (BAR)

Goodall (1989): "Loud, sharp sounds, usually given in long sequences with much variation in pitch." Functions to protest to another individual of same or different species (e.g. baboons). Probably corresponds to bonobos' Bark (Okayasu 1991) or Wa! call (Kano 1998). See also <cough bark>, <pant-bark>, <waa bark>, and <wraa>. (Type A) Category 2. See video.

bark eating
See <eat phloem>.

bark, infantile (BAI)
High-pitched bark emitted by infants. (Type A) Category 2–3. See video.

battle (BAT)
Direct intergroup agonistic interactions, in which both sides contain multiple males (Manson and Wrangham 1991; Wilson and Wrangham 2003). Two groups concertedly attack each other. (Type B) Category 1–3.

beat chest (BCH)
Hit chest with fists. Idiosyncratic element of male charging display at Gombe (Goodall 1971). Cf. <slap self>. Absent at Mahale. (Type A).

beat ground with stick
See <club ground>.

beat sole with palm (BSP)
Lie supine and repeatedly and simultaneously beat each sole with palm of corresponding left or right hand. Play solicitation signal and also solo play shown by adolescent male, Cadmus. (Type A) Category 8–9. See video.

beat with fist
See <hit>.

bed
Goodall (1989) Nest: "Platform or bed for sleeping or resting." See also <make day bed>, <make night bed>, <grunt in bed>, and <play in bed>.

beg (BEG)
Goodall (1989): "Beg for food, toy, or any desirable object from the possessor. Includes <extend hand to beg> and <mouth for begging>. Begging is often accompanied by whimpering, and if unsuccessful, the beggar may even throw a tantrum." Kano's (1998) Food beg for bonobos. (Type D) Category 1. See video. See also video "reject-move."

beg with hand
See <extend hand to beg>.

beg with mouth
See <mouth for begging>.

bend and release (BER)
Whiten et al. (1999) Branch din: Bend and release saplings, so noisy movement alerts others. Customary only at Lopé. Recorded among Lomako bonobos during inter-group encounters (Hohmann and Fruth 2003). Absent at Mahale. (Type A).

bend and stretch knee (BAS)

Sit orthograde, bend and stretch knee while extending arm palm down. Shown by adult male, Fanana. Idiosyncratic greeting response to adult female. (Type A) Category 9. See video.

bend away (BEN)

Goodall (1989): "With elbow and wrist flexed, arm drawn close to body, the chimp leans slightly away from a passing higher ranking animal. This is a submissive gesture." van Hooff's (1973) Flinch/shrink. Plooij's (1984) Bend away (BEN). Also seen in bonobos of Wamba (Kano 1998). (Type A) Category 1–2.

bend branch (BEB)

Bend attached branch when making bed or cushion. See <make bed> and <make cushion>. (Type A) Category 2. See video.

bend shrub (BES)

Bend shrub, grass, bamboo or herb with one or both hands. In making day-bed or in courtship on ground, done in sitting or quadrupedal posture. In charging display done bipedally. Done also by immatures as solo play. (Type A) Category 3. See video.

bend shrub in courtship (BEC)

Nishida (1997) Shrub-bend courtship. Male courtship pattern at Mahale. While in sitting or quadrupedal posture, push down or pull stem of shrub, bamboo grass, or herb such as ginger, and put one foot on plant; repeat many times same series of actions, thus making crude ground-cushion or bed. Usually followed by <stamp> or <rap> ground, or rarely by <hop quadrupedal on spot >. Customary among males at Mahale. Seen also as courtship at Taï (Boesch 2009). Absent in bonobos of Wamba. (Type B) Category 7. See video.

Beni

Study area of *Pan troglodytes schweinfurthii* in eastern province of DRC (0°N, 30°E). Research done by A. Kortlandt in 1960, 1963 and 1964 (Kortlandt 1962, 1996).

bipedal hunch

See <hunch bipedal>.

bipedal jump

See <leap bipedal> and <hop bipedal on spot >.

bipedal run

See <run bipedal>.

bipedal sex dance

See <dance bipedal>.

bipedal swagger

See <swagger bipedal>.

bipedal transport
See <transport bipedal>.

bipedal walk
See <walk bipedal>.

birth
See <give birth>.

bite (BIT)
Use teeth to apply force to object. Includes Goodall's (1989) Bite: "nipping or cutting into the anatomy of another individual by pressing the teeth on the skin and closing the jaws hard." Plooij's (1984) Bite (BIT). Goodall (1989) did not differentiate from <mouth>. Cf. <mouth> and <play bite>. Bonobo females bite more often than males (Kano 1998). Male bonobos do not bite females (de Waal and Lanting 1997). (Type A) Category 1.
See video.

bite and pull hairs (BIP)
Player bites some hairs of partner and pulls them with teeth. Infant male, Ichiro, once bit and pulled, and another time sucked, hairs of adult male. Juvenile male, Xmas, bit and pulled hair of back of infant in play. Playful behavior. (Type B) Category 8–9. See video.

bite self (BIS)
Bite own palm, wrist, or arm. Pattern expressed in temper tantrum, in response to stressful situation. (Type A) Category 3. See video.

bluff over
See <hunch over>.

bob (BOB)
Goodall (1989): "The body bobs up and down as elbows are flexed and straightened. Shown typically by adolescent males but also by adolescent females, juvenile males and adult males when a high-ranking individual passes. Sometimes as the dominant recipient of the gesture moves on, the bobber backs away in front of the dominant animal. This may provoke an aggressive response. Bobbing is accompanied by pant-grunts which may become rather frenzied pant-screams." Plooij's (1984) Bob (BOB). van Hooff's (1973) Squat bob. Absent in bonobos (Kano unpublished). (Type A) Category 3.
See video.

bob bipedal (BBB)
Like <bob>, but standing on feet only. (Type A) Category 3–6. See video.

boredom
Goodall (1989) Boredom. See <kill time>.

Bossou

Study area of *Pan troglodytes verus* in southwestern part of Guinea (7°39′N, 8°30′W). Research organised by Y. Sugiyama and T. Matsuzawa, continuing from 1976 to present (Sugiyama 2008). Pioneer studies done by A. Kortlandt in the early 1960s and his colleagues in 1968–1969 (Albrecht and Dunnett 1971).

bounce foot/feet up and down rhythmically (BFU)

Lie supine, flex and extend leg rhythmically, thus bounce foot up and down. Element of <balance>. (Type A) Category 1. See video.

bow (BOW)

Plooij (1984) Bow (BOW): Similar to bob in context. Deep flexion of legs but arms extended forward, so that head is lower than hips. Absent in bonobos (Kano unpublished), but see Pika (2007, p. 49). (Type A) Category 3. See video.

brachiate (BRA)

Goodall (1989) Brachiation. Hanging from branch, swings hand-over-hand to move to terminal branch of tree to eat or to descend to ground. Plooij's (1984) Brachiate (BRA). Same as Kano's (1998) Brachiation and Susman et al.'s (1980) Armswinging. Cf. <swing>. (Type A) Category 1. See video.

branch break

See <break branch>.

branch-clasp

See <groom-branch-clasp>.

branch din

See <din branch>.

branch drag

See <drag branch>.

branch-flail

See <flail long object>.

branch-haul

See <hook branch>.

branch-hook

See <hook branch>.

branch shake

See <shake branch>.

branch slap

See <slap branch>.

break bone at joint (BBJ)
Hold and snap limb bone at joint with both hands. Seen in <eat meat>. Cf. <twist>. (Type A) Category 1–7. See video.

break branch (BRB)
Chimpanzee detaches branch and throws or drags it in display on ground (see <drag branch>), or eats from it comfortably above ground. In feeding, chimpanzee breaks off terminal branch laden with fruits or young leaves, carries it to safe place, such as big bough or position closer to tree trunk, and eats from branch. Many large food trees therefore modified into different shapes. Recorded for *P. t. troglodytes* of Gabon (Takenoshita et al. 1998). Seen in bonobos in feeding and before branch-dragging display (Kano 1998). (Type A) Category 1–2. See video.

break branch with foot (BRF)
Before eating *Crematogaster* ants or *Xylocopa* bees, hold both ends of thick, dry branch by hand and use foot to push forcibly against it, in order to break it in two. Also reported from Goualougo (Sanz and Morgan 2007). (Type A) Category 1–7. See video.

break open by pushing chin (BOC)
Hold large fruit with fairly soft shell such as ripe *Saba comorensis* or lemon (*Citrus limon*) in mouth and push chin with hand or wrist in order to break open and eat contents. (Type A) Category 3–6.

break tree (BRT)
Break tree trunk by leaping and pushing against it, or pushing and pulling with hand repeatedly. Kortlandt's (1967) Break tree. Element of charging display by adult male. <leap>, <push>, <pull>. Also when eating pith from small tree. Also in bonobos (Kano unpublished). (Type B) Category 2. See video.

break wind
See <fart>.

bristle (BRS)
Goodall (1989) Hair erection: "The hairs stand on end." Plooij's (1984) Hair erection or pilo erection (HAI). Bonobos bristle, but less so (Kano 1998). (Type A) Category 2. See video.

brush aside (BRU)
Mother sweeps off infant her back by hand in refusing to carry it. Male sweeps away by hand infant who wedges itself between copulating pair. Kano's (1998) Rebuff for bonobos. (Type A) Category 1.

brush away from branch (BBA)
Brush repeatedly with hand dry branch to wipe away ants on it. Used to remove *Crematogaster* ants, when eating their pupae, larvae, or eggs. (Type A) Category 7. See video.

brush away from self (BSF)

Rubs away ants with hand which cling to its own body. (Type B) Category 7. See video.

brush-tip fishing probe (BTF)

Modified probe used for termite fishing. "Removes the leaf from the distal end of the tool with his/her mouth. One end of the tool is then pulled sideways through partially closed teeth (most often their molars) several times to fray the end of the herb into a brush tip. Some individuals were also observed biting the end of the tool with their incisors to pull apart the fibers" (Sanz et al. 2004). First described as "brush-stick" by Sugiyama (1985) for chimpanzees of Campo. Absent at Mahale. (Type B) See video.

Budongo

Study area of *Pan troglodytes schweinfurthii* in Budongo Forest Reserve, Uganda (1°44′N, 31°33E). Research organized by V. Reynolds, continuing from 1990 to present. Pioneer studies done by V. and F. Reynolds (1965) in 1962–1963, by Y. Sugiyama (1968) in 1966–1969, and A. Suzuki (1971) in 1969–1970.

bump

See <pass with body contact>.

butt

See <push head into ventral>.

buttress-beat

See <drum>.

Bwindi

Study area of *Pan troglodytes schweinfurthii* in Bwindi Impenetrable National Park, Uganda (0°53′–1°08′S, 29°35–29°5′0E). Research organised by C. Stanford from 1996 to 2005 (Stanford and Nkurungi 2003, Stanford 2008).

C

C

call

See <vocalize>.

Campo

Study area of *Pan troglodytes troglodytes* within Campo Animal Reserve, Cameroon (2°30′N, 9°30′E). Research done by Y. Sugiyama in 1984–1985 (Sugiyama 1985).

cannibalism

See <eat infant>.

care alloparentally (CAL)

Nishida (1983a) Alloparental behavior: Class of behavioral patterns similar to maternal care, shown to infants by individuals other than mother, including <hug>, <groom>, <transport>, <mouth>, <lie and hug>, <put dorsal>, <pat>, <touch>, <balance>, <protect>, <scratch>, <share food>, <take finger in mouth>, etc. Similar behavior in bonobos (Kano 1998). (Type D) Category 1. See video. See also video "retrieve infant."

care alloparentally for another species (CAA)

Alloparental behavior directed to member of another species. Adolescent female carried dead leopard cub to day bed, hugging and grooming it (Hiraiwa-Hasegawa et al. 1986). Cf. <trifle with>. (Type D) Category 8–9. See Fig. 6.

C

Fig. 6 Care alloparentally for another species: Juvenile female carried dead leopard cub into tree and groomed it (By courtesy of Richard Byrne)

care maternally (CAM)

Complex of nurturing behavior shown by mothers to their offspring. Consists of <hug>, <suckle>, <transport>, <groom>, <protect>, <play socially>, <share food>, <share bed>, <reject>, <support>, etc. No juvenile offspring is suckled, transported, or allowed to share bed. Adolescent offspring still groomed by mother but usually on reciprocal basis, and supported occasionally during fighting with another individual. Mother grooms adult offspring reciprocally. Bonobo mother supports her adult son even against his rival (Kano 1992). (Type D) Category 1. See video.

carry

See <transport>.

catch with hand (CAT)

Seize flying insect such as fly by quick sideways jerking movement of hand, then open palm to inspect it. Also in bonobos (Kano unpublished). <grab> is used for other contexts. Cf. <whisk fly with arm>. (Type A) Category 2. See video.

catch with hands (CAB)

Clutch food thrown by human. When chimpanzees of K group were given bananas and sugarcane, some chimpanzees of both sexes skillful at catching piece of sugar cane with both hands, even when in tree. Behavior developed without training, in response to human throwing to chimpanzees. (Type A) Category 8.

change nipple (CHN)

During sucking, infant changes nipple from right to left or vice versa (Nishida 1993a). Plooij's (1984) Change nipple (CHA). (Type D) Category 1. See video.

charge (CHG)

Goodall (1989): "Fast run directed toward another individual." Bonobos charge too. (Type C) Category 2. See video.

charging display

See <display, charging>.

chase (CHA)

Goodall (1989) Chase. Run after fleeing individual, seeking to grab it in aggression, play or hunt. Chaser usually dominant to chased, but not always. Chasing play between adults is slower. Goodall (1989) "Sometimes, a subordinate individual, screaming or perhaps uttering waa barks, chases after his/her aggressor as the latter charges away" See <aggress morally>. See also <circle orthograde> or <circle quadrupedal>. Bonobo chasing is more ritualized, as chaser often gets ahead of fleeing individual (Kano 1998). Kano (1998) refers to "Chase play" for chase in play context. (Type C) Category 1. See video.

check penile erection (CPE)

Estrous female approaches male in order to investigate whether he is ready to copulate. Nishida (1997): Erection check. (Type C) Category 8–9. See video.

chest-beating

See <slap self>.

chew (CHE)

Plooij (1984) Chew (CHE). Press item repeatedly between upper and lower teeth. Also occur in bonobos. (Type A) Category 1. See video.

childhood

See Juvenile. "Childhood" should be applied only to human children, who are not suckled by mother but yet cannot obtain food enough to subsist by themselves.

choke in tantrum (CHT)
Goodall (1989) Glottal cramps: “During loud and prolonged tantrum screaming, the chimpanzee may seem to choke and only hoarse squeaks or rasping sounds can be heard.” See also <throw temper tantrum>. Absent in bonobos? (Type A) Category 3–5.

circle orthograde (CIO)
Occurs in similar context as <circle quadrupedal>. Individual moves round and round tree bipedally or tripedally, while holding tree trunk with hand(s) and keeping torso upright. See also <circle quadrupedal>. (Type A) Category 1–5. See video.

circle orthograde off ground (CIF)
Hold tree trunk by hand while revolving around it with feet off ground. Solo play or solicit play. (Type A) Category 3–9. See video.

circle quadrupedal (CIQ)
In social play, two or more individuals may chase round and round tree, another individual, or without any pivot-point. One or both of players may transport leafy twig in mouth. Chased one may take turn to chase, and vice versa. Occasionally occurs as solo play. Often accompanied by play-pant at Mahale and Gombe. Seen in bonobos of Wamba (Enomoto 1997). See also <circle orthograde>. (Type A) Category 1–2. See videos.

clack teeth (CTE)
Goodall (1989) Teeth-clack: “Mouth rhythmically opened and closed during grooming. The teeth are clapped together each time the mouth is closed. Usually the lips are quite tightly drawn over the teeth when the mouth is open. Occurs during vigorous grooming.” See also <sputter>, and <smack lip>. Not seen in bonobos of Wamba (Kano 1998). (Type A) Category 3–5. See video.

clamp elbow (CLE)
Goodall (1989) Elbow clamp: “When small infant pesters mother for food, she may clamp it to her breast with both elbows and continue her meal over its head.” See also <fend>. Cf. <cover nipple>. (Type A) Category 3–5.

clap hand (CLP)
Koops and Matsuzawa (2006) Hand clapping. Unhabituated adult female of Nimba Mts. once observed to clap hands as part of threatening gesture against human observers. Only record of hand clapping by wild chimpanzees, although known in captivity. Absent at Mahale. (Type A)

clasp self (CLS)
Goodall (1989) Self-clasp: “Sit or lie with arms tightly clasped around self... in a relaxed manner during resting.” Not seen at Mahale, nor in bonobos of Wamba (Kano 1998). Infants of both species of *Pan* in captivity show this gesture. (Type A)

clear nasal passage
See <probe nasal passage>.

climb (CLI)
Plooij (1984) Climb (CLI). Includes not only <climb vertical> and <climb vertical, extended elbow>, but also climbing by <walk quadrupedal on palms> on slanting trunk or bough of big tree. Goodall's (1989) Vertical climb ("The chimp climbs up a trunk, branch, pole, or other vertical structure") and Kano's (1998) "Vertical climb up" for bonobos probably includes both <climb vertical> and <climb vertical, extended elbow> (Type C) Category 1. See video.

climb cliff (CLR)
Similar to human "rock-climbing" without any artificial climbing equipment. (Type B) Category 1–3. See video.

climb down
See <descend>.

climb vertical (CLV)
Hunt et al.'s (1996) "Flexed-elbow vertical climb". Climb woody vine or small tree by alternately using palmar surfaces of four limbs. (Type A) Category 1. See video.

climb vertical, extended elbow (CVE)
Hunt et al. (1996) "Extended-elbow vertical climbing". Extend arms around trunk of large tree and propel upwards by simultaneous thrusting or walking movements of feet. (Type A) Category 1. See video.

cling (CLN)
Goodall (1989): "Gripping a tree trunk or branch with hands or hands and feet. An infant clings to mother's or other's belly, back, arm or leg..." Also in bonobos (Kano unpublished). (Type A) Category 2. See video.

clip leaf by hand (CLH)
Tear leaf by hand, separating midrib and leaf-blade. At Mahale, adolescent male, Cadmus, used to solicit mating. Customary at Budongo, and habitual at Taï and Kibale (Whiten et al. 1999). (Type A) Category 9. See video.

clip leaf by mouth (CLL)
Nishida (1980b) Leaf clip: Pull leaf repeatedly between lips or teeth with one hand, producing conspicuous sound that attracts attention from prospective sex partner. Among most common courtship displays at Mahale. Act of frustration or play at Bossou (Sugiyama 1981) and prelude to make drumming-enacted frustration at Taï (Boesch 1995). Thus, contexts differ from place to place, e.g., nuances of behavior differ between Mahale and Taï. Customary at Budongo and Ngogo. Reported as habitual solicitation of social play for Lomako bonobos (Hohmann and Fruth 2003). See also <use midrib for fishing probe>. (Type B) Category 2. See video.

C

clip leaf for fishing probe (CLM)
Use midrib of leaf as fishing probe after removing leaf-blade by leaf-clipping with mouth or hand. Customary in fishing for carpenter ants at Mahale (Nishida 1973) and termite-fishing at Assirik (McGrew et al. 1979). (Type B) Category 7. See video.

close eyes (CEY)
Close eyes: (1) in sleep, (2) as reflex when frightened, and (3) when seemingly in bliss. Adult male, Alofu, lies on ground with eyes closed, apparently in bliss, after stuffing many sweet fruits in his mouth. Plooij's (1984) Eyes Closed (EYC). (Type A) Category 1. See video.

club (CLB)
Strike target with long (usually more than 1 m), thick stick in overarm movement. Directed to conspecific, another species, or inanimate object. van Hooff (1973) and Kortlandt and Kooij (1963) described this pattern. Seen rarely in a few individuals at Mahale. Habitual at Taï and Gombe (Whiten et al. 1999). Present at Bossou and Assirik. Absent in bonobos (Kano unpublished). (Type A) Category 3–8. See video.

club ground (CLG)
Strike ground with long, thick stick to solicit copulation, by two juvenile males. Adult male, Fanana, clubs ground to coerce estrous female to follow him during consortship. Juvenile male clubs the ground to threaten another juvenile during play or to threaten newly-immigrated female and human observer. Several adult males, Lukaja, Fanana, and Dogura, club ground or wall of house as part of charging display. (Type A) Category 8. See video.

club other (CLO)
Club another individual with stick. Juvenile males, Michio and Cadmus, club another in solicitation of play. Adult male, Toshibo, twice hit another adult male with stout stick after flailing it during fighting. (Type A) Category 3–8. See video. See also video "fight"

coalition
Long-term and strategic mutual support against common rival. Bonding made by pair such as Pinky and her babysitter, Gwekulo, and Opal and her adult daughter, Ruby. Cf. <alliance>. See also <attack concertedly>. See video.

coalitionary attack
See <attack concertedly>.

coalitionary mate guarding
See <herd in coalition>.

collect dry leaves (CDL)
Infant or juvenile reclines or sits and collects dry leaves by raking, pulling or grabbing and hugs them in solo play. (Type B) Category 7. See video.

comb (CMB)

Whiten et al. (1999) Comb: Use stem to comb through hair. Present at Budongo and Goualougo (Sanz and Morgan 2007). Absent at Mahale. (Type A)

community

See <unit group>.

compress lips (COM)

Goodall (1989) Compressed lips: "The lips are pressed tightly together so that the upper lip is bunched up and protrudes beyond the point where the lips meet. Seen during displays and attacks." Plooij's (1984) Compressed lips (COM). Absent in bonobos. (Type A) Category 1–3. See video.

confiscate (COF)

To prevent aggression, take away potentially dangerous object such as rock or stout branch from another who is irritated. At Arnhem Zoo, adult female confiscated stone from adult male (de Waal 1982). De Waal (2009)'s Disarm. At Mahale, however, only mothers took away stone or stick from adolescent male or juvenile offspring. (Type D) Category 1–3. See video.

conflict behavior

Situational acts, when two or more contradictory drives are stimulated in actor, such as <groom carcass>, <kick up>, <shake hand side by side quickly>, etc.

consolation

See <console>.

console (CSL)

de Waal and van Roosmalen (1979) Consolation. Third party touches, hugs, grooms or licks victim attacked and wounded by another. Consolation reported from Gombe (Goodall 1971), Mahale and Taï (Boesch 2009). See <lick wound>. Kano's (1998) "Conciliating behavior" by bonobos. (Type D) Category 1. See video.

consort (CST)

Goodall (1989) Consortship: "An exclusive sexual relationship between adult male and female. ...the female is taken to a peripheral part of the community range away from other rival males." McGinnis's (1973) Consortship, Tutin's (1975) Safari behavior. Consortship seen for Taï chimpanzees (Boesch and Boesch-Achermann 2000). At Mahale, consortship is mating strategy of non-alpha males. See also <lead>. Cf. <herd>. Not seen in bonobos of Wamba (Kano 1998). (Type D) Category 3. See Fig. 7.

C

Fig. 7 Consort: Adult male, Kamemanfu and young estrous female, Wakapala of K group, in consort visited Myako camp to eat sugarcane when main party was absent (T. Nishida)

contact water (WAC)
Contact with water includes: <stir water>, <look at water>, <drink>, <play with water>, <splash water>, <throw splash>, etc. See also <wade>. (Type D) Category 1–7.

contact with nipple (CNP)
Weanling puts mouth to mother's nipple without sucking. Final stage of weaning. (Type A) Category 1. See video.

coprophagy
See <eat feces>.

copulate (COP)
Intromission and pelvic thrusting between male and estrous female. Goodall's (1989) Copulation, Plooij's (1984) Copulate. (Type A) Category 1. See video. See also video in "bend shrub in courtship", "check penile erection", "dart", "erect penis", "press teeth against back", and "wedge".

copulate between group (CPG)
Young, and less often prime, estrous females visit neighboring unit group and copulate at Gombe (Goodall 1986), Mahale (Nishida et al. 1985) and Taï (Vigilant et al. 2001). Bonobo unit groups occasionally mingle peacefully, and copulation by members of different groups occurs at Wamba (Idani 1990) and Lomako (Fruth, cited in de Waal and Lanting 1997). Adult

males do not interfere with such copulations (Kano 1998). (Type C) Category 1.

copulate dorso-ventral (CDV)
Male mounts female from rear. Major copulation pattern of mature males and females in chimpanzees and bonobos. (Type A) Category 1. See video. See also video "interfere in copulation."

copulate ventro-ventral (CVV)
Male mounts female from front. Occurs rarely in adolescent females and immature male chimpanzees. More common in adolescent females and mature and immature male bonobos (Kano 1998). (Type A) Category 1.

copulation pant
See <pant in copulation>.

copulation scream
See <squeal in copulation>.

copulatory dart
See <dart>.

copulatory squeal
See <squeal in copulation>.

core area
Area used intensively by individual chimpanzee although also used by other members of unit group (see Hasegawa 1990). Females compete one another for good core area (Kahlenberg et al. 2008; Pusey et al. 2008). Cf. <territory>.

cough (COH)
Goodall (1989):"When chimps have colds and coughs they sneeze and cough, sounding like humans." See <sickness>. Bonobos also cough (Kano unpublished). (Type A) Category 1. See video.

cough bark (COB)
Goodall (1989): "Soft bark. A grunt-like sound uttered through slightly open mouth directed by higher ranking chimpanzees to subordinates. Indicates mild annoyance. It functions as a mild threat." Threat call of one syllable. Also in bonobos (Kano unpublished). (Type A) Category 2. See video.

cough-threat
See <cough bark>.

counter-attack
See <retaliate>.

courtship display
See <solicit copulation>.

cover breast
See <cover nipple>.

cover from rain (COR)
Cover body with leafy twig to avoid rain. Present at Goualougo (Sanz and Morgan 2007). Also seen for bonobos of Wamba (Kano 1982b). Cf. <shelter from rain>. Absent at Mahale. (Type B)

cover nipple (CON)
Plooij (1984) Cover nipple (CON). Goodall's (1989) Cover breast: "Mother prevents access to breast by placing her arm across her nipple...She may also lie with breasts pressed to the ground or branch". Part of mother's weaning of infant. Absent in bonobos of Wamba (Kano 1998). (Type A) Category 3. See video.

cover self (CVS)
Cover self with dry leaves or sand. Solo play of infants. (Type A) Category 8. See video.

cradle with hand (CDH)
Mother supports newborn infant with arms. Goodall's (1989) Cradle. Cf. <transport with hand support>. Kano's (1998) Cradle for bonobos of Wamba. (Type A) Category 1.

cradle with leg (CWL)
Mother hangs from tree branch supports infant clinging to her belly with one thigh flexed, so that baby's body is pressed against her. Cf. <transport with thigh support>. Plooij's (1984) Cradle (CRA). Seen in bonobos (Kano unpublished). (Type A) Category 2.

creep (CRE)
Walk stealthily. Goodall (1989) states that "..a fearful subordinate tries to creep from the vicinity of a particular higher ranking individual. The subordinate proceeds with extreme caution, moving slowly while keeping a constant sharp watch on the other..." Bonobos of Wamba show similar pattern (Kano 1998). (Type C) Category 2. See video.

cross (CRS)
Flex one or both limbs to rest on opposite limb. (Type B) Category 1. See video.

cross arms (CRA)
Flex arms across chest. (Type A) Category 1. See video.

cross arms on head (CAH)
Flex arms on head. (Type A) Category 3–7. See video.

cross legs (CRL)
Flex one leg to rest on opposite leg, which is supported by substrate. (Type A) Category 1. See video.

crouch (CRO)

van Hooff (1973) and Plooij (1984) Crouch (CRO). Goodall's (1989) Crouch: "Quadrupedal posture with limbs flexed, hindquarters not turned towards another animal". Goodall states that: "...Often occurs in greeting when it is usually accompanied by pant grunts or during submission after aggression..." Kano's (1998) Crouch and Present in agonistic interaction for bonobo. (Type A) Category 2. See video.

crouch-present

See <present with limbs flexed>.

crowd (CRW)

Many chimpanzees gather to peer at individual or gaze at object. Occurs when chimpanzee tries to extract thorn from foot, inspects seriously injured skin, carries newly-dead baby, possesses prized food such as meat, or an unusual object in environment that stimulates curiosity. When crowded, chimpanzees appear to allow closer contact with each other than usual. No adult bonobos crowd (Kano unpublished). (Type D) Category 1. See video.

crush (CRH)

Press small arthropod such as louse between finger-tips. Absent in bonobos. (Type A) Category 3–8. See video.

crutch (CRU)

Goodall (1989): "A method of progressing, usually when the chimp is going downhill, during which the arms are used as crutches. With both hands on the ground, the legs and body are swung forward through the arms..." Plooij's (1984) crutch (CRU). At Mahale, also occurs when a mother carries newborn on belly, accompanied by <transport with thigh support>. In bonobos, crutch used only as last step of a long walk or as one-step movement while eating or resting (Kano 1998). (Type A) Category 2. See video.

cry

See <whimper-scream>.

culture

Behavior learned socially after birth, shared by at least most members of one age/sex class in group. See also <tradition> and <fashion>.

cushion

See <make cushion>.

D

D

dab (DAB)

Goodall (1989): "Rapid hitting movement in which the backs of the flexed fingers are directed toward a more dominant individual, typically an adult male. Only a few individuals, mostly adolescents, have been observed to dab. Usually these individuals dabbed several times in succession, and the gesture was mostly directed toward the face of the other." Plooij's (1984) Dab (DAB). At Mahale, seen only in few adolescents and juveniles when pant-grunting to adult males. Absent in bonobos of Wamba (Kano 1998). (Type A) Category 8.

dab leaf (DAL)

Whiten et al. (1999) Leaf-dab: Dab leaf on wound and examine it. Customary at Kibale, present at Gombe, Goualougo and Taï. Absent at Mahale. (Type A).

dance orthograde (DCB)

Nishida (1997) Bipedal sex dance: Stand bipedal while raising arm(s) or performing branch-shaking display. Accompanied by <scratch self>. Done by estrous female irritated by apparent indifference of adult male to female's sexual advance. Only two adult females, Wakilufya and Ako, of M group showed this type of courtship, but de Waal (1982) mentioned similar pattern. Similar pattern in bonobos of Wamba. (Type B) Category 8.

dangle (DAN)

Plooij (1984) Dangle (DAN). Goodall (1989): "Infant hangs, under, from the side of, or below mother or another individual with one or both hands. Does not grip with feet. May also dangle from a branch…" See <hang>. Seen in bonobos of Wamba (Kano 1998). See also <ride dangling>. (Type A) Category 2. See video.

dart (DAR)

Estrous female runs quickly few meters after copulation. Often accompanied by copulatory squeal and grin, then, female often lies down for few minutes. Nishida's (1997) Copulatory dart. Absent in bonobos (Kano unpublished). (Type A) Category 3–5. See video.

day nest
See <make day bed>.

death
Chimpanzees die of diseases, predation, mother's death, fall from tree, killing by conspecifics, injuries, etc. (Nishida et al. 2003). See also <die>.

deceive (DCV)
(1) Mislead others to own advantage by not showing posture, gesture, or facial expression usually done in that context, (2) mislead others to own advantage by showing posture, gesture, or facial expression that is normally inappropriate or irrelevant in that context (Nishida 1990, 1998). Reported from Gombe (Goodall 1971) and captivity (de Waal 1982). (Type D) Category 1. See video.

deep incursion
See <intrude>.

defecate (DFC)
Excrete feces. See also <diarrhea> and <push peri-anogenital region with finger>. (Type C) Category 1. See video. See also video "push peri-anogenital region with finger".

defecate, prone (DFL)
Lie prone and defecate, when chimpanzee has diarrhea. (Type A) Category 3–8. See video.

defecate quadrupedal (DFQ)
Stand on four limbs and defecate. Common posture. (Type A) Category 2.

defecate, sitting (DFS)
Defecate while crouching with weight on legs only. Common posture of defecation. (Type A) Category 1. See video.

defecate with hand support
See <push peri-anogenital region with finger>.

depart (DPT)
Begin to travel, increasing distance from starting point. (Type C) Category 1. See video.

depart together (DPG)
Mother approaches infant to initiate travel together. Older infants and juveniles often precede mothers, who then may follow them. See also <pull>, <put dorsal>, <put ventral>, <scoop infant>, <solicit riding>, <solicit companion>, and <grasp and push shoulders>. Cf. <go ahead>. (Type C) Category 1. See video.

dermatophytosis
Infections skin disease on face of chimpanzee (Nishida, Fujita, Matsusaka et al. 2007). See video.

descend (DES)

Climb down from elevated site: Includes six patterns: <descend by brachiation>, <descend tree trunk feet first>, <descend tree trunk head first>, <leap down>, <slide down vertically> and <descend orthograde feet first>. Cf. <fall>. (Type C) Category 1. See video.

descend by brachiating (DBR)

While progressing by brachiation, supportive bough gradually bends down until ground can be reached with feet. Also in bonobos (Kano 1998). (Type A) Category 2. See video.

descend orthograde feet first (DFF)

Grasp branch below with feet to move to another tree below. (Type A) Category 1. See video.

descend tree trunk feet first (DQF)

Climb down backwards. Seen in bonobos (Kano unpublished). (Type A) Category 1. See videos.

descend tree trunk head first (DQH)

Climb down forwards. Seen in bonobos (Kano unpublished). (Type A) Category 2. See video.

detach

See <kidnap>.

detain (DEN)

Hold body part of partner such as leg in order to continue social play, etc. while partner tries to escape. Mother restricts infant's movement e.g., by grabbing its legs when it approaches possibly dangerous object such as adult male. (Type C) Category 1. See video.

detour (DET)

Deliberately take roundabout route. When chimpanzee seeks to avoid dominant rival, or when shortest route is difficult, e.g. gap in tree or crossing river. Kano's (1998) Detour and Pass with detour for bonobos of Wamba. (Type C) Category 1. See video.

diarrhea (DIR)

Loose stools often seen after eating fruits of *Pycnanthus* or *Toddalia*. Also when being attacked by another, or traveling near territorial boundary. See <sickness>. (Type A) Category 1.

die (DIE)

Lie motionless and breathless. Dead body usually found on ground in bush along animal track. See <death>. (Type D) Category 1. See Fig. 8.

dig (DGG)

Make hole in ground using hands or stick. (Type B) Category 1.

Fig. 8 Die: Infant of Wakasunga died of disease (T. Nishida)

D

dig for army ant with hand/stick (DGA)
Chimpanzees of Bossou dig up army ants' (*Dorylus*) nest by hand but also use digging stick (Sugiyama 1995a). Absent at Mahale. (Type A).

dig for underground storage organ by hand (DGH)
Dig up root such as *Aeschynomene* sp. tree by hand while sitting or standing quadrupedally. Chimpanzees of Tongo dig up tubers as substitute for drinking (Lanjouw 2002). Cassava is dug up and eaten at Bossou (Hockings et al. 2010). (Type A) Category 3–6. See video.

dig for underground storage organ with tools (DUS)
Dig up underground storage organs with tool, reported only from Ugalla (Hernandez-Aguilar et al. 2007). Absent at Mahale. (Type A).

dig for water by hand (DGW)
Dig well in wet ground of dry streambed using one or both hands to get drinking water. At least five individuals of M group do this. Behavior common at Semliki (Hunt and McGrew 2002). Bonobos of Lilungu and Wamba do not dig for water (Bermejo et al. 1994; Kano 1998). See also <remove objects from water surface>. (Type A) Category 3–5.

dig for water with stick (DGS)
Use tool to dig wet ground of dry streambed to get water. Only one juvenile female of Mahale seen to do this. Absent in bonobos of Wamba (Kano unpublished). (Type A) Category 8–9.

dig stingless bees' nest with stick (DGB)

Customary at Gashaka (Fowler and Sommer 2007). Also present at Bwindi (Stanford et al. 2000). Absent at Mahale. (Type A).

dig termite nest with stick (DGT)

Dig into termite nest with stout stick. Known from Ndoki (Suzuki et al. 1995) and Goualougo (Sanz et al. 2004). Absent at Mahale. (Type A).

dip ant and wipe (DPW)

Ant-dip (McGrew 1974), Ant-dip-wipe (Whiten et al. 1999). Dip wand of vegetation into massed driver ants, and when ants climb wand, remove ants with sweeping movement of loosely flexed hand and pop ants into mouth. Custom at Gombe, Assirik, Gashaka (Fowler and Sommer 2007), and Goualougo (Sanz and Morgan 2007). Present at Bossou (Humle and Matsuzawa 2002). Likely to occur at Kalinzu (Hashimoto et al. 2000), Fongoli (McGrew et al. 2005) and Ngotto Forest (Hicks et al. 2005). Absent at Mahale. (Type B) See Fig. 9.

dip ant directly (DPD)

Whiten et al. (1999) Ant-dip-single. Chimps dip short stick (rod) into masses of driver ants and when ants climb stick, quickly mouth them off stick. Custom at

Fig. 9 Dip ant and wipe: Adult female, Skosha, of Gombe dips (**a**) and wipes (**b**) *Dorylus* ants (By courtesy of Caroline Tutin)

Fig. 9 (continued)

Bossou, Taï and Goualougo (Sanz and Morgan 2007). Present at Gombe and Nimba (Sugiyama 1995a). Absent at Mahale. (Type B) See video (Bossou).

dip fluid (DPF)

Whiten et al. (1999) Fluid dip: Use probe to extract fluids such as honey. Some immature chimpanzees use sticks for drinking water in tree hollows (Matsusaka et al. 2006). See also <leaf-sponge>. Customary at Gombe, Mahale, Kibale, Lopé, Assirik and Taï. Present at Goualougo (Sanz and Morgan 2007). (Type A) Category 3–8. See video.

dip fruit wadge into water (DFW)

Chimpanzees of Taï dip fruits and their wadges of *Sacoglottis gabonensis* into water apparently to suck out most of sweet juice (Boesch 2009). Absent at Mahale. (Type A).

D

dip hand and lick water (DPL)

Immature chimpanzee immerses hand into stream or crevice of tree and lick water from fingers, knuckles, or back of hand (Matsusaka et al. 2006). (Type B) Category 3–8. See video.

directed scratch

See <scratch self>.

disarm

See <confiscate>.

discard (DIC)

Abandon object. (Type C) Category 1. See video.

discard fruit skin with mouth (DIF)

Remove fruit skin with movements of lips, teeth and tongue, then abandon it by mouth. (Type A) Category 1. See video.

disperse seed (DIS)

Excrete feces away from place where eaten fruits originated, so chimpanzees unintentionally disperse seeds. See <swallow seed>. (Type D) Category 1. See Fig. 10.

Fig. 10 Disperse seed: Seeds of fruit recovered from chimpanzee feces (T. Nishida)

display as contest (DSC)

When male does charging display, more dominant male rushes up and displays against him. May elicit displays by more dominant males, and chain may continue up to five individuals. If final displayer is alpha male, display contest ends. (Type C) Category 3–7. See video.

display away (DSA)
Goodall (1989): "After interaction with another, usually an attack, the chimp departs in a charging display." Seen in bonobos (Kano unpublished). (Type C) Category 2.

display, charging (DPC)
Plooij (1984) Charging display (CHD). Goodall (1989)'s description of Charging display is modified here. May move in slow rhythmic gait, run at moderate speed or very fast. Display patterns include <scratch dead leaves>, <drag branch>, <shake branch and sway woody vegetation>, <bristle>, <compress lips>, <slap>, <stamp>, <slap-stamp>, <flail>, <throw>, <drum>, <rake>, <swagger, bipedal> and <pant-hoot>. May be bipedal, quadrupedal, or tripedal. Typically male display though some females do it also. Goodall (1989) discriminated Non-vocal and Vocal displays. The former when displayer makes no calls, tends to be directed toward another individual, and may be followed by <attack>. It is typical component of male dominance rivalry. Vocal display is accompanied by <pant-hoot> and typically not directed toward any other chimpanzee. Common during reunions, food excitement and when individual has been frustrated in obtaining desired goal. At Gombe, charging display is sometimes followed by <beat chest>, but not at Mahale. Bonobo's charging display is accompanied by <charge>, <stamp>, and sometimes by <drag branch> (Kano 1998). (Type C) Category 2. See video.

display past (DSP)
Goodall (1989): "When a chimp, who is clearly directing a display toward another, passes close, and may hit or kick the other in passing." In 1995 at Mahale, two adult males regularly hit human observers during charging displays. Seen in bonobos. (Type C) Category 2–5.

display, rain (DSR)
Goodall (1989) Rain display or Rain dance. At Mahale, two types of displays occur at onset or during heavy rain or storms. Not directed at other individuals. One is a noisy, active display accompanied by <pant-hoot>, <charge>, <run bipedal>, <stamp> etc. Other is prolonged, slow tempo display, of which main components are <sway woody vegetation>, <shake branch>, <run bipedal>, <climb rapidly>, <brachiate> and <leap>. Customary at Gombe, Mahale, Kibale, and habitual at Taï and Budongo. At start of storms or heavy rain, adult male bonobos may run few meters while calling loudly, but no other activities such as those of chimpanzees occur (Kano 1998). (Type C) Category 2–3. See video.

display, streambed (DSS)
Charging display done in water course. Goodall's (1989) Stream bed displays resemble rain displays but occur in streambeds. Goodall's (1989) Waterfall display probably should be included here. At Mahale, main components of streambed display are <run>, <leap>, <drag branch>, <throw splash>,

<shake branch>, <swing> and <sway woody vegetation>. Absent in bonobos (Kano unpublished). (Type C) Category 3–5. See video.

display toward (DST)

Goodall (1989): "Display directed toward one or more others." Seen in bonobos (Kano unpublished). (Type C) Category 2. See video.

distance scratch

See <scratch self distantly>.

distract (DTR)

Shift another's attention away from what it tries to do. For example, mother grooms or plays with infant to divert its attention from sucking from nipples during weaning. If infant is male and throws temper tantrums in response to his mother's refusal, she sometimes allows him to copulate with her. Goodall's (1989) Distract. Also in bonobos of Wamba (Kano 1998). (Type D) Category 2. See video.

distress call (DCL)

Includes <hoo>, <staccato call>, <whimper>, and <whimper-scream>. (Type D) Category 2–3. See video.

Dja

Study area of *Pan troglodytes troglodytes* in Dja Biosphere Reserve of southeast Cameroon (3°37′–3°45′N, 13°11′–13°17′E). Research done by Deblauwe and colleagues from 2002 to 2005 (Deblauwe et al. 2006).

dominant

Higher-ranking of two individuals. See also <pant-grunt>. See Fig. 11.

Fig. 11 Dominant: Subordinate male (*right*) showed temper tantrum when dominant male approached (T. Nishida)

dorso-ventral embrace
See <mount>.

drag and circle (DRC)
Older individual holds arm or hand of playmate in mouth and drags other in social play (<drag by mouth> below). Performer may circle around tree. (Type B) Category 3–5. See video.

drag branch (DRB)
Goodall (1989) Branch dragging: "A branch broken from a tree, or one lying on the ground is dragged by one hand during a charging display…" Finally, branch is typically thrown forward. Widely known across Africa: Ndoki (Kuroda 1998), Bossou (Sugiyama 1998), Taï, Lopé, Gombe, Mahale, Kibale, Budongo (Whiten et al. 1999) and Goualougo (Sanz and Morgan 2007). In bonobos of Wamba, branch-dragging displays (Ingmanson 1996; Kano 1990, 1998) have many more behavioral elements than those of chimpanzees. Late adolescent male bonobos use display to tease and to challenge males higher in rank. Bonobos sometimes drag branch while walking without stamping or vocalizing. Bonobos of Lomako also drag branch (Hohmann and Fruth 2003). (Type A) Category 2. See video.

drag by hand (DRA)
Goodall (1989): Drag by hand along ground individual, branch, palm frond, large food object, etc. Shown in aggression, threat, play, and maternal solicitude. Plooij's (1984) Drag (DRA). Bonobos of Wamba drag only branches (Kano 1998). (Type A) Category 1.

drag by mouth (DRM)
Drag while holding object, animal or another individual in mouth. Cf. <pull with mouth>. (Type A) Category 2–3.

drag carcass by hand (DCH)
Drag large carcass by hand along ground, moving forward. Simple transport or element of charging display. (Type A) Category 3–5. See video.

drag carcass by mouth (DCM)
Drag large carcass forward by mouth along ground. Simple transport. (Type A) Category 3–5.

drag dry leaves (DDL)
Walk forward fast quadrupedally with head down and shoulders hunched, while pushing dry leaves with hands and feet, producing distinctive sound. Adult males do in charging displays. Adolescent female, Ivana, teased newcomer female, Qanat, by displays including this. Done by juvenile as solicitation of play. Cf. <push leaf-pile>. (Type A) Category 3–7. See video.

drag other by hand (DRS)
Goodall (1989) Drag other. Drag by arm and leg in aggression or play. Mother often drags older infant when infant delays departing with her.

Adult male may abuse infant by dragging it forcibly along path. Kano's (1998) Drag other and Pika's (2007) Grab-push-pull for bonobos. (Type A) Category 1. See video.

drag other by mouth (DOM)

Drag another individual while holding hand, arm, foot, leg, nape or back in mouth in play. Usually move backwards. (Type A) Category 2–3. See video. See also video "hold body part in mouth."

drag to kill (DRK)

Adult male, Musa, once ran and dragged with one hand large red colobus monkey that he had captured, so that monkey was hit and injured by woody vegetation and choked by woody vines, causing death. Tactic prevented monkey from biting chimpanzee predator. (Type A) Category 3–8. See videos.

drape (DPE)

Goodall (1989) Drape: Drape bunch of fruits, leaves, or skin of colobus over shoulders, nape or head (McGrew and Marchant 1998). (Type A) Category 1–5. See video.

drink (DRI)

Goodall (1989): "Drinking of water or other liquids. This includes drinking directly by leaning over and sucking from the water source in a quadrupedal or sitting posture; dripping water from the fingers into the mouth; dipping fingers into water & licking them; drinking with or licking back of hand, licking drops of water from surfaces such as leaves, hair, and so on…" See also <dig for water>, <keep water in mouth>, <leaf-sponge>, <leaf-spoon>, and <remove objects from water surface>. Bonobos rarely drink from running or standing water (Kano 1998). (Type B) Category 1. See video. See also videos "leaf-spoon"and "remove objects from water surface."

drink from hole in tree (DRH)

Put hand, fingers, or leaf sponges into hollow of tree, then lick water extracted. Customary at Gombe (Goodall 1968), Mahale (Matsusaka et al. 2006) and Bossou (Tonooka 2001). (Type B) Category 3–5. See video.

drink from lake (DRL)

Chimpanzees of lakeside groups, B, K, M, N, and L groups, drink from Lake Tanganyika. Some afraid of high waves and do not do so. (Type A) Category 1–6. See video.

drink from stream (DRR)

Chimpanzees of Mahale mostly depend on streams and rivers for drinking water. (Type A) Category 1. See video.

drip (DRP)

Juvenile male, Xmas, once dribbled water repeatedly from back of hand at pool. Solo play with water. (Type A) Category 8–9. See video.

drop (DRO)

Let object fall, occasionally deliberately onto another individual or human observer. (Type B) Category 1. See video.

drop bark bits (DBS)

Remove by hand without obvious purpose outer bark little by little and drop. Shown by nervous individual or when waiting for companion engaged in long bout of sleep, grooming, or fishing for ants or termites, etc. See also <kill time>. (Type A) Category 1–6.

drop branch (DPB)

Drop branch from overhead to threaten human observer (or occasionally another conspecific) below, sometimes hitting target. Some adult males forcibly kick dead branches, detaching them, so that they fall. Occasionally seen at Mahale when chimpanzees were not well habituated. Seen in female chimpanzee in Gabon (Takenoshita et al. 1998). Ingmanson (1996) described "Object dropping", e.g. branch in bonobos of Wamba. (Type A) Category 2. See video.

drop infant (DPI)

Mother drops older infant when mother tired during long bout of travel. Includes <shake rump>, <shrug>, <lower head and shoulder>, <reject-sit>, <push>, <lower rump>, etc. (Type B) Category 3–5. See video. See also video "lower rump."

drop self (DPS)

Release branch and let self drop to lower vegetation or ground without leaping. Common practice when leaving tree. Youngsters in play repeat sequence of climbing and dropping. Occasionally fall occurs accidentally when branch breaks. Cf. <leap down> and <fall>. (Type A) Category 1. See video.

drum (DRU)

Component of charging display. Goodall (1989): "Hit and/or kick tree trunks, especially those with buttresses. Often they grab the buttress with their hands and stamp on it with their feet. They may kick backwards with their feet or hit the trunk or buttress with their hands. Often a particular "drumming tree" triggers a drumming display and the individuals of a traveling group are likely to drum one after the other as they pass. Usually drumming is accompanied by the pant-hoot call, but sometimes … drumming is without vocalization. Chimps at Gombe also pound on the walls of research buildings and on a barrel placed at the feeding area …." Plooij (1984) Drum (DRU). Mahale chimpanzees do same. Drumming patterns vary individually. For example, one male stands erect and beats walls of research buildings with only one palm, another does so with two palms, third one with one fist, fourth one kicks with one foot, and fifth one pounds with both feet (Nishida 2003b). See also <kick> and <slap>.

Drumming reported from Kibale and Taï (Arcadi et al. 1998). Bonobos drum on buttresses (Kano 1998; Hohmann and Fruth 2003), most often immediately after onset of traveling (Kano 1998). Cf. <beat chest>. (Type C) Category 2. See video.

drum belly

See <slap self>.

dunk (DUN)

Immerse object in water without releasing it. Juvenile male, Michio, once dunked fruit in puddle, then sucked it (Matsusaka et al. 2006). Adult male, Musa, dunked colobus skin in stream apparently to clean it (Nishida 1993b). Wadge dipping observed at Taï (Boesch 1991b). Absent in bonobos of Wamba (Kano unpublished). <leaf-sponge> is kind of dunking. See <wash>. (Type B) Category 1–5.

dunk face (DUF)

Juveniles repeatedly submerge face in river in play. (Type A) Category 8. See video.

E

eat (EAT)

Remove foodstuff (leaves etc.) from substrate directly with mouth or by hand, <bite>, <chew>, and <swallow> it. Sometimes, <process food> or <wadge> before swallowing. Goodall's (1989) Feed. Kano's (1998) Feed. (Type B) Category 1. See video.

eat algae (ELG)

Enter stream and eat algae, standing quadrupedal in water up to 30 cm deep. Adolescent immigrant female, Sally, only few times seen to eat. Likely customary in groups inhabiting higher elevations of Mahale (Sakamaki 1998; Nishida et al. 2009). Also recorded at Bossou (see <scoop algae>) and Odzala National Park (Devos et al. 2002). (Type B) Category 8. See videos (Mahale and Bossou).

eat ant (EAA)

Ants are eaten everywhere, but species eaten vary from place to place (McGrew 1992). For example, army ants (*Dorylus*) are not eaten at Mahale, but at many other sites (Schoening et al. 2007). Bonobos of Wamba do not eat ants, but Lilungu bonobos eat adult and larvae of *Tetraponera* (Bermejo et al. 1994). Humans in the tropics eat ants (Bodenheimer 1951). See <eat *Camponotus* ant>, <eat *Crematogaster* ant>, and <eat *Oecophylla* ant>. (Type B) Category 1. See video. See also videos "brush away from branch", "brush away from self" and "fish for carpenter ants."

eat beetle larva (EBL)

Ingest larvae of Coleoptera. Occasionally seen at Mahale. Regularly eaten at Bossou (Sugiyama 2008). Likely to be eaten at Ngogo (Sherrow 2005). (Type B) Category 1–6. See video.

eat blossom (EAB)

Ingestion techniques resemble those for eating leaves. Eaten by bonobos (Kano 1992). (Type B) Category 1. See video.

eat *Camponotus* ant (ECA)

Eat carpenter ants (*Camponotus* spp.) by cracking open grass stem or by using tools (see <fish for carpenter ant>). (Type B) Category 1–7. See video.

eat carcass (EAX)

Scavenge. Eat carcass of mammals, such as red colobus, red-tailed monkey, bushbuck, blue duiker, etc., not killed by self or one's companions. Scavenging occasionally seen at Gombe (Muller et al. 1995) and Mahale (Hasegawa et al. 1983; Nishida 1994), but virtually absent at Ngogo (Watts 2008a). Absent in bonobos of Wamba. (Type B) Category 1–3.

eat *Crematogaster* ant (ECR)

Eat *Crematogaster* ants (*Crematogaster* spp.) by splitting open grass stem or dry branch with teeth, breaking off large branch with hands or occasionally hands and feet. See <break branch>, <break branch with foot> and <brush away from branch>. (Type B) Category 1–7. See video.

eat Dorylus ant (EDO)

Dorylus ants are available, but never eaten at Mahale. See also <dig for army ant with hand/stick>, <dip ant and wipe> and <dip ant directly>. (Type B) See video (Bossou).

eat egg (EAE)

Eggs of birds such as francolin (*Francolinus squamatus*) and guinea fowl (*Guttera edouardi*) eaten by chimpanzees of Mahale and Gombe. Gombe's chimpanzees readily ate domestic fowl eggs when offered (Goodall 1971). Some Mahale chimpanzees also ate them. (Type B) Category 1–5. See video.

eat eye mucus (EEM)

Eat discharge from eyes. Rare. (Type B) Category 8–9. See video.

eat feces (EAD)

Goodall (1989) Coprophagy. At Mahale, eating feces occurs rarely. Adult female, Chausiku, picked out and ate undigested seeds of *Saba comorensis* from her feces. Gombe apes picked undigested meat from feces (Wrangham 1977). Assirik chimpanzees picked out baobab seeds from feces (McGrew et al. 1988). Semliki apes defecated directly into their own hand and raised feces to mouth to reingest seeds of *Saba comorensis* (Payne et al. 2009). Bonobos of Wamba reingested seeds of *Dialium* (Sakamaki 2010). (Type B) Category 2–8.

eat fruit, inner skin (EFI)

Scrape and ingest inner skin (endocarp) of fruit such as *Saba comorensis*. (Type B) Category 1–6. See video.

eat fruit, pulp (EAF)

Pick fruits in tree, from shrubs or herbs on ground, or pick up fallen fruits from ground, then, open up fruit shell with hands or mouth, and ingest contents. (Type B) Category 1. See video. See also video "spit seed."

eat gall (EAG)
Gall of insects (Diptera) inhabiting parts of plants such as *Milicia*. (Type B) Category 3–5. See video.

eat honey (EHO)
Insert finger or probe into honeycomb and eat honey of honeybees (*Apis*), or gnaw with incisors to widen entrance of nest of stingless bees (*Trigona*). Honey eating widespread in chimpanzees (McGrew 1992; Stanford et al. 2000; Hicks et al. 2005; Sanz and Morgan 2009). Bonobos of Lilungu eat honey of stingless bees (Bermejo et al. 1994). (Type B) Category 1.

eat infant (EAC)
Cannibalism. Chimpanzees of Mahale occasionally eat infants (Kawanaka 1981; Hamai et al. 1992). Known also from Budongo (Suzuki 1971), Gombe (Bygott 1971; Goodall 1977), Ngogo (Watts and Mitani 2000) and Taï (Boesch and Boesch-Achermann 2000). Absent in bonobos (Kano 1992). (Type B) Category 3. See Fig. 12.

Fig. 12 Eat infant: Alpha male, Ntologi, ate infant of newly-immigrated female (By courtesy of Tamotsu Asou)

eat insect (EIN)
Chimpanzees of Mahale eat ants, honey bees, stingless bees, termites, and beetle larvae (Nishida and Hiraiwa 1982). See also <eat ant> and <eat termite> for details. (Type D) Category 1.

eat invertebrate
See <eat insect>. Chimpanzees eat only insects, but bonobos eat earthworms.

eat leaf (EAL)

Put leaves into mouth and ingest them. Bring mouth to leaves or bring plucked leaves to mouth. Important cropping technique is <pull through>. (Type B) Category 1. See video. See also videos “pull through with mouth” and “store”.

eat meat (EAM)

Ingest teeth, bones, bone marrow, brain, entrails, stomach and intestinal contents, and skin, in addition to muscle. Eating meat unusually prolonged, sometimes for more than 3 h. Prey include red colobus, blue, red-tailed, and vervet monkeys, yellow baboon, greater galago, infant chimpanzee (see <eat infant>), blue duiker, bushbuck, bushpig, giant rat, rock hyrax, civet, mongoose, sun squirrel, francolin, guinea fowl, chicken, weaver bird, and black kite (Takahata et al. 1984; Hosaka et al. 2001; Fujimoto and Shimada 2008). Meat-eating occurs throughout chimpanzee range (Uehara 1997). After Uehara’s review, black and white colobus, white-cheeked mangabey and red duiker (Watts and Mitani 2002) and banded mongoose (Bogart et al. 2008) recorded as prey. Bonobos eat meat of blue, bay, and black-fronted duikers, flying squirrel, fruit bat, galago, red-tailed monkey, black mangabey, Wolf’s guenon, rodent, birds, etc. Recorded at Wamba (Kano 1992), Lomako (Hohmann and Fruth 1993), Lilungu (Sabater Pi et al. 1993; Bermejo et al. 1994), and Lui Kotale (Hohmann and Fruth 2007; Surbeck and Hohmann 2008), but less often and extensively (Kano 1998). See also <hunt>. (Type B) Category 1. See video.

eat nasal mucus (EAU)

Eat nasal mucus after picking nose or sniffling. Adult male often eats nasal mucus of his partner during social grooming. See also <pick nose>, <probe nasal passage>, <sneeze> and <sniffle>. (Type B) Category 1. See video. See also video “sneeze.”

eat *Oecophylla* ant (EOE)

Grab weaver ants’ nest (*Oecophylla longinoda*) with one hand, peel it apart on ground, quickly eating ants. (Type B) Category 1–7. See video.

eat petiole (ETP)

Eat only petiole (and discard other leafy parts) of *Ipomoea rubens*, *Cordia millenii*, *Myrianthus arboreus*, *Vernonia amygdalina*, etc. (Type B) Category 3–5. See video.

eat phloem (EAH)

Eat phloem or inner bark rich in sugar and protein by scraping inner surface of outer bark with incisors. Eaten widely by humans (Nishida 1976). (Type B) Category 1. See Fig. 13.

Fig. 13 Eat phloem: Inner bark of *Brachystegia bussei* teeth-marked by chimpanzees of K group (T. Nishida)

eat pith (EAP)

Eat soft, juicy piths of herbs such as elephant grass, gingers, ginger lilies, and *Marantochloa*, herb vines such as *Ipomoea*, woody vines such as *Landolphia* and *Saba*, and shrubs such as *Vernonia* by removing outer surface with teeth. Pith of *Megaphrynium* eaten by both bonobos and humans at Wamba (Kano 1992). (Type B) Category 1. See video. See also video "peel with teeth."

eat resin (EAR)

Scrape resin of *Terminalia mollis* with incisor or pick large lump of resin of *Albizia glaberrima* with fingers. Bonobos (Kano unpublished) and African pastoralists eat resin. (Type B) Category 1. See video.

eat rock (EAK)

Nibble or lick rock. See <lick rock>. (Type B) Category 6.

eat root (ERT)

Dig ground to reach and eat live root of tree of *Aeschynomene* sp. Cf. <dig>. (Type B) Category 7. See video.

eat seed (EAS)

Crack seed pod with teeth, chew and ingest seeds of plants such as *Pterocarpus*, *Diplorhynchus*, *Baphia*, *Parkia*, etc. (Nishida and Uehara 1983). Bonobos eat *Pterocarpus tinctorius* (Kano 1992). (Type B) Category 1–2. See video. See also video "beg."

eat semen (EAN)

Females, more often than males, eat semen ejaculated in copulation. (Type B) Category 3–7. See video.

eat termite (ETT)

Topple tower of termite mound with hands and pick up soldiers or winged reproductive forms of termites (*Pseudacanthotermes*). Fishing for soldiers of *Pseudacanthotermes* recorded only for K group chimpanzees (Uehara 1982). Fishing for soldiers of *Macrotermes* has been seen in chimpanzees of B group (Nishida and Uehara 1980, McGrew and Collins 1985). Widely known throughout chimpanzee range (McGrew 1992). See <fish for termite> and <shake wet arm to catch termite>. Bonobos of Lilungu eat termites of different genera from those eaten by chimpanzees by inserting fingers into hole opened manually (Bermejo et al. 1994). Termites eaten widely by tropical people (Bodenheimer 1951). (Type B) Category 1. See videos (Mahale and Goualougo).

E

eat termite soil (ETS)

Pick up and ingest small pieces of soil from termite towers of *Pseudacanthotermes*. Humans and bonobos (Kano 1998) eat termite soil. (Type B) Category 1. See video.

eat vertebrate

See <eat meat>.

eat vomit

See <reingest vomit>.

eat with foot (ETF)

Pick up and put food into mouth with foot. Infant male, Caesar, plucked several ripe fruits of *Garcinia* with foot and ingested them. (Type B) Category 8–9. See video.

eat wood

See <eat xylem>.

eat xylem (EAW)

Eat dry dead wood of *Pycnanthus angolensis*, *Ficus vallis-choudae*, *Garcinia huillensis*, etc. Unknown elsewhere. (Type B) Category 6. See video.

Ebo Forest

Study area of *Pan troglodytes vellerosus* in Ebo Forest, Cameroon (4°6′N, 10°23′E). Research organized by B.J. Morgan and E.E. Abwe from 2005 to present.

ejaculate (EIA)

Ejection of seminal fluid from penis. Males begin to ejaculate at 9 years old at Mahale. Semen often eaten by both sexes. See <eat semen>. (Type A) Category 1. See video.

elbow clamp

See <clamp elbow>.

emasculate (EMA)

Remove testicles, occasionally in extreme male-male competition. Evidence from Kanyawara (Muller 2002) and Taï (Boesch 2009), as well as from captivity (de Waal 1998). Some male bonobos of Wamba lack testicles but cause unknown (Kano 1984). (Type D) Category 1–3.

embrace full (EMF)

Goodall (1989) Ventro-ventral embrace: "Two individuals face each other and each puts one or both arms around the other." van Hooff's (1973) Embrace, Plooij's (1984) Embrace full (EMF). Contexts include greeting, reassurance, alliance, and reconciliation. Embrace full by adult males may be accompanied by <thrust>. In bonobos, embrace full posture seen only in genito-genital rubbing by females and ventro-ventral copulation between males and females (Kano 1998), so both embrace-full in erect posture and embrace-full by adult males are lacking. (Type A) Category 1–3. See video. See also video "console".

embrace half (EMH)

Plooij (1984) Embrace half (EMH). Goodall's (1989) Arm round: "Puts one arm around another as in a half embrace." Often occurs when one individual pant-grunts to another. Adult male puts his arm around another (usually dominant) male while walking and pant hooting. Occurs in copulatory solicitation in bonobos (Kano unpublished). (Type A) Category 1. See video.

emigrate (EMI)

Adolescent female disperses from natal unit group and enters another group for reproduction, at about 11 years old. Parous females rarely leave resident group. See also <transfer>. (Type D) Category 1.

enter hole (ENH)

Enter cavity, such as burrow of aardvark in exploration or play in juveniles. (Type C) Category 1. See video.

erect hair

See <bristle>.

erect penis (EPE)

Plooij (1984) PEN. Goodall's (1989) Penile erection. Occurs during sexual and food excitement. See also <adduct penis>. Kano's (1998) Penile erection for bonobos. (Type A) Category 1. See video. See also video "open thighs."

escape (ESC)

Leave area when more dominant individual targets present in party.
For example, when estrous female moves to another sex partner, she departs while often looking back or glancing at other adult males to monitor possible attack. (Type D) Category 1–3.

estrous cycle

See <swelling of sexual skin>.

estrus

Female copulation confined mostly to period of maximal swelling of sexual skin, which includes day of ovulation. See <swelling of sexual skin>.
See Fig. 14.

Fig. 14 Estrus: Sexual skin with maximal swelling (T. Nishida)

expel (EXP)

Whiten et al. (1999) Expel/stir. Nishida's (1973) Expelling stick. Push stout stick into hole of tree or other cavity and stir it violently side to side or back and force with arm. Mahale chimpanzees try to expel rock hyraxes from cavities in rocks (Nakamura and Itoh 2008). Customary at Gombe, Mahale and Taï. (Type B) Category 1–3. See video.

extend (EXT)

Extend arm or leg to another individual. (Type B) Category 1. See video.

extend arm as ladder (EXA)

Nishida (1983a): Caregiver extends arm upwards in tree, to allow infant to climb up easily, by making ladder with arms. See also <aid in locomotion> and <extend leg as ladder>. Absent in bonobos of Wamba. (Type C) Category 3–7. See video.

extend hand (EXH)

Goodall (1989) Extend hand. van Hooff (1973) Hold out hand: "A variety of forms were observed. Most common is a form in which the actor, either sitting or standing, extends its arm roughly horizontally towards a fellow. The arm is in a position about midway between pronation and supination (i.e. with the thumb up). The hand may be bent at the wrist so that its back is turned to the partner with the fingers bent or fully stretched…" Goodall's (1989) "Extend hand" limited to gesture with wrist and fingers extended and palm up or down. Goodall (1989) described varying contexts in begging, as reassurance to out-of-reach subordinate, and when submissive chimpanzee seeks reassurance after aggression, or solicits support from third party. In mother-infant interactions, extend hand is contact-seeking. For bonobos, extend hand seen only as <extend hand, palm downward> or <extend hand, palm upward> below in food begging or mother seeking contact with offspring (Kano 1998). (Type B) Category 1. See video. See also video "extend."

extend hand and put knuckle on ground (EKG)
Put arm forward and place knuckle on ground (instead of holding tree trunk). Often performed to solicit grooming as in <extend hand to hold tree>. Seen in bonobos (Kano unpublished). (Type A) Category 2.

extend hand, palm downward (EXD)
Goodall (1989) Arm stretch: "The arm or arms are extended toward another. The palm of the hand usually faces down. Seen in courtship or greeting." Also as solicitation of play. Mother may invite infant to riding on her or to go ahead. van Hooff (1973) Stretch over, Plooij (1984) Extend hand, palm downward (EHD), Pika (2007) Reach, Nishida (1997) Arm-stretch. Seen in bonobos (Kano unpublished). (Type A) Category 2. See video.

extend hand, palm sideways (EXS)
Arm extended to another with elbow bent, palm sideways, and fingers straightened or flexed, occasionally accompanied by panting. Greeting and when mother seeks to retrieve infant from rough social play. Function may be reassurance, appeasement, or reconciliation. (Type A) Category 3–7. See video.

extend hand, palm upward (EXU)
Plooij (1984) Extend hand, palm upward (EHU). Bipedal or quadrupedal. Used to solicit play, to beg, when mother solicits infant to ride on her belly or back, or to pant-grunt to dominant individual. Once, juvenile female, Flavia reached left arm to TN while standing bipedally. Seen in bonobos (Kano 1998). (Type A) Category 1. See video. See also video "pant-grunt with bent elbow."

extend hand to beg (EXB)
Plooij (1984) Beg with hand (BWH): Stretch hand to possessor's hand, mouth or food. Goodall's (1989) Beg hand-to-food, hand-to-hand and hand-to-mouth. Kano's (1998) Food beg: hand to hand and hand to mouth. Cf. <mouth for begging>. (Type B) Category 1. See video.

extend hand to hold tree (EHT)
Reach arm forward and grasp tree trunk. Often performed to solicit grooming. Seen in bonobos. (Type A) Category 2.

extend leg (EXL)
Reach leg instead of hand to another individual, with similar function to <extend hand>. Shown especially when performing <extend hand>

difficult, e.g. in tree or when performer holds infant and food. Seen in bonobos. (Type A) Category 2. See video.

extend leg as ladder (ELL)

Mother extends leg to help infant to come to her. See also <aid in locomotion> and <extend arm as ladder>. (Type A) Category 3–6. See video.

extended grunt

See <grunt, extended>.

extract (EXR)

Sanz and Morgan (2007) "Gathering an item or fluid with a tool for the purpose of extracting it from a location that is difficult to access manually." Includes <fish>, <dip>, <pick out bone marrow>, etc. (Type B). Category 1–3.

eyes closed

See <close eyes>.

eyes open

See <open eyes>.

F

F

face close

See <peer>.

fall (FAL)

Drop by accident from higher level to lower level. Cf. <drop self> and <leap down>. (Type D) Category 1. See video.

fall over backward (FDB)

Topple backward from seated position, arms extended. Thud caused by hitting dry leaves; sometimes after being pushed. Signals play solicitation as attention getter. Weanling uses to attract mother's attention. Cf. <somersault, backwards> and <headstand>. (Type A) Category 3–5. See video.

family

Chimpanzee family consists of mother and her immature offspring and adult sons. If daughters do not emigrate, it may become group of three or more generations. See Fig. 15.

Fig. 15 Family: Old adult female, Wabunengwa, and her adolescent daughter and infant son (T. Nishida)

fantasy play
See <play, imaginary>.

farsightedness
See <presbyopia>.

fart (FAR)
Break wind. (Type A) Category 1. See video.

fashion
Behavior learned socially, shared by at least most members of at least one age or sex class, and transmitted horizontally within one generation. See also <culture> and <tradition>.

F

feed
See <eat>.

female solicit
See <solicit copulation>.

fend (FEN)
Goodall (1989): "Most often seen when a mother keeps her child away with her hand or foot when she is being pestered for a share of her food, when her infant tries to suckle during weaning, and so on. But adults sometimes fend off individuals who are begging from them." Not limited to food but any case of deflecting reach of another, e.g., fend against being tickled. See also <clamp elbow> and <cover nipple>. Also in bonobos (Kano 1998). (Type D) Category 1. See video.

fight (FIG)
Mutual attack. See <attack>. (Type D) Category 1. See video.

Filabanga
Study area of *Pan troglodytes schweinfurthii* in western Tanzania (5°24′S, 30°10′E). Research done from 1965 to 1966 by T. Kano and colleagues (Kano 1972).

fill mouth with food (FMF)
Stuff food by hand into mouth. (Type A) Category 1. See video.

finger in mouth
See <push finger into mouth> and <take finger in mouth>.

finger wrestling
See <wrestle with fingers>.

fireman slide
See <slide down vertically>.

fish (FIS)
Insert flexible strip of bark, vine, twig, grass etc. into nest of social insects, withdraw and pick off insects with lips, teeth or tongue. Absent in bonobos

of Wamba, Lomako, and Lui Kotale (McGrew 2004). (Type B) Category 1–3. See video.

fish for army ant

See <dip ant and wipe> and <dip ant directly>.

fish for carpenter ant (FIA)

Insert fishing probe such as peeled bark, vine, twig, modified branch, midrib of leaf, or split wood into entrance of wood-boring carpenter ants'nest (*Camponotus* spp.), withdraw probe laden with soldiers, and take them into mouth with lips and tongue. Usually arboreal and aseasonal, sometimes continues for more than 3 h (Nishida and Hiraiwa 1982). Customary at Assirik, Lopé, and Gashaka (Fowler and Sommer 2007). Chimpanzees of Kasakela group at Gombe do not fish for carpenter ants, but those of Mitumba do. (Type B) Category 3–5. See video. See also videos "fish", "supplant", "take", "use tool" and "wait turn."

fish for termite (FIT)

Resembles ant fishing. Fishing for the termites of *Pseudacanthotermes spiniger* at Mahale only in K group (Uehara 1982, Fig. 16). Fishing for *Macrotermes* only in B group (Nishida and Uehara 1980; McGrew and

Fig. 16 Fish for termite: Termites of *Pseudacanthotermes* were fished for by chimpanzees of K group (By courtesy of Shigeo Uehara)

Collins 1985). Fishing for termites not seen in M group, apparently because *Macrotermes* is absent in range. Recorded for all subspecies of chimpanzees; Mt. Assirik (McGrew et al. 1979), Bossou (Humle 1999), Fongoli (McGrew et al. 2005; Bogart and Pruetz 2008), Gashaka (Fowler and Sommer 2007), Campo (Sugiyama 1985), Ndoki (Suzuki et al. 1995), Goualougo (Sanz et al. 2004; Sanz and Morgan 2007) and Belinga (McGrew and Rogers 1983), in addition to Gombe. Not recorded for bonobos. Bantu people fish for termites with more elaborate techniques (Bodenheimer 1951). (Type B) Category 3. See videos (Goualougo).

fission and fusion

Chimpanzee unit groups (communities) repeatedly split into small parties, then later rejoin into larger ones, depending largely on abundance and distribution of food supply. Also bonobo unit groups and human hunting bands.

fist

See <hit>.

fist ground

See <hit ground with fist>.

fixed stare

See <stare fixedly>.

flabby bottom

Goodall (1989): “Loose and wrinkled sexual swelling during the final stage of detumescence.”

flail (FLL)

Brandish arm or branch for threat display, solicitation of play, or solo play, or brandish and beat animal to ground in order to kill it. See also <flail arm> or <flail long object>. (Type B) Category 1. See video.

flail arm (FLA)

Brandish or bring down arm as threat display or play solicitation. (Type A) Category 1–2. See video.

flail long object (FLB)

Brandish branch or other long object. Goodall (1989): Brandish branch in hand and “waves this weapon at an opponent.” Plooij’s (1984) FLL. Kortlandt’s (1967) Tree-swish. At Mahale flailing at another rarely done by adult male. Flailing at human observer more rarely done by adolescents. Adult male of Mahale caught, flailed and hit to death monkey or bird against ground. Cf. <shake branch>. Branch-flailing is element of branch-dragging among bonobos of Wamba (Kano 1998). (Type A) Category 1–2. See video.

flap (FLP)

Goodall (1989): “…downward slapping movement of the hand, usually repeated several times, in the direction of another individual. Often seen in female squabbles…" Plooij's (1984) Flap (FLP). Absent in bonobos of Wamba (Kano 1998). (Type A) Category 3–5.

flee (FLE)
Goodall (1989) Flight (Run away): "rapid progression away from an alarming or dangerous stimulus…" Also seen in chasing play. Elements include <scream>, <run>, <climb>, <jump>, and <descend>. van Hooff's (1973) Flight. Kano's (1998) Flee. (Type C) Category 1. See video.

flee after startle (FLS)
Goodall (1989) Startle flee: "When a resting chimp is suddenly startled by an unexplained sound he or she may run to climb the nearest tree, or at least hold the trunk ready to climb. One chimp in a resting group who behaves thus will trigger the same reaction in most or all of the others." Also in bonobos (Kano 1998). (Type B) Category 2. See video.

flee from colobus male (FLC)
Adult male colobus occasionally descends to ground and chases chimpanzees away. Both adult males and females fearful of colobus male and run away. Also chimpanzees of Gombe, but those of Taï do not flee from colobus (Boesch 2009). (Type D) Category 5–7. See video.

flight
See <flee>.

flinch
See <bend away> and <parry>.

flip lip (FLI)
Goodall (1989) Lip flip: "The upper lip is rolled up and back towards the nose. Usually relaxed situation…" Plooij's (1984) FLI. Function of everted lip facial expression unknown, but may reflect boredom. Absent in bonobos (Kano unpublished). (Type A) Category 3–5. See video.

flop (FLO)
Lie supine and kick, rake, and wriggle on ground in pile of dry leaves, like flopping fish on line. Solo play of juvenile male, Xmas, perhaps idiosyncratic to him. (Type A) Category 8. See video.

fluid-dip
See <dip fluid>.

fly whisk
See <whisk fly with leafy stick>.

fold leaf (FLD)
Tonooka (2001) Leaf-folding: Tear leaf or leaves with hand or mouth, stuff into mouth, chew them repeatedly, take them out of mouth and hold between index and middle fingers, soak them in water in hollow, pick up and suck water from them. Final shape is not like sponge, but folded leaves. Cf. <leaf-sponge>. Only at Bossou? Absent at Mahale. (Type B) See Fig. 17.

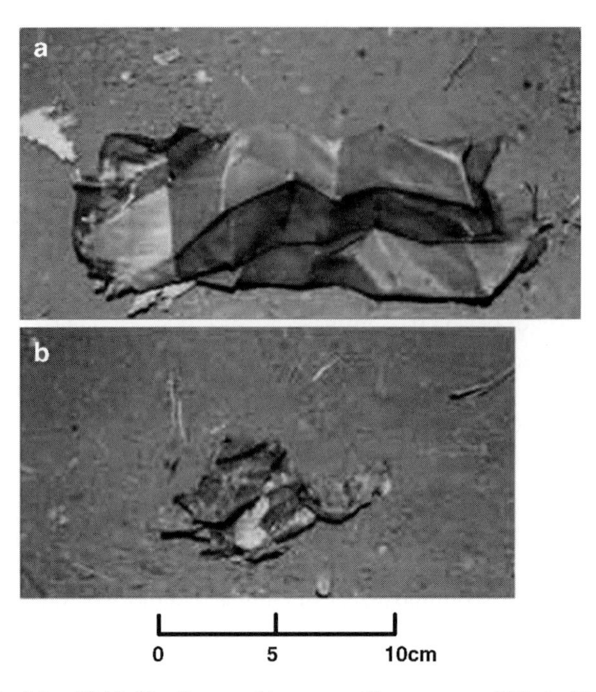

Fig. 17 Fold leaf: Leaf folded by Bossou chimpanzee (By courtesy of Rikako Tonooka)

follow (FOL)

Trail after, trying to maintain close proximity (but not physical contact) to another. Goodall's (1989) Follow. Plooij's (1984) Follow. Kano's (1998) Follow. (Type C) Category 1. See videos.

follow in contact (FOC)

Follow another with hands placed on back or shoulder of leader. This occurs when follower feels uneasy, is in stressful situation, enlists support from leader, or is interested in sexual swelling of leader. (Type B) Category 3–6. See video.

follow specific female (FSF)

Newly-immigrated female chooses, follows, and forms affiliative relationship with one particular resident female by grooming her and caring for her infant (Nishida 1989). Also in immigrant female bonobos (Furuichi 1989; Idani 1991). (Type C) Category 2. See video.

fondle penis

See <fumble penis>.

Fongoli

Study area of *Pan troglodytes verus* in southeastern Senegal (12°40′N, 12°12′W). Research organized by J. Pruetz from 2001 to present.

food grunt

See <grunt, food>.

food-pound on other

See <pound food on other>.

food-pound on wood

See <pound food on wood>.

food scream

See <scream, food>.

food sharing

See <share food>.

food transfer

See <share food>.

friendship

Long-term association with mutual dependence. Cf. <coalition>. Past-prime male, Kalunde and past-prime female, Nkombo, traveled and rested most of time together, and she supported Kalunde when he was challenged for alpha status by adult male. See video.

fumble clitoris (FMC)

Adult or adolescent female repeatedly touches own clitoris. Males occasionally grasp clitoris of young females. Rarely seen. (Type A) Category 1–8. See video.

fumble nipple (FMN)

Manipulate own nipples with thumb or finger for self-reassurance or self-stimulation. Done by few chimpanzees (one male and three females in 1997). Occurs when frightened or nervous when dominant individuals closeby (Marchant and McGrew 1999). Usually thumb only or thumb and index finger are used. Adult male, Alofu, dexterously alternates between right and left nipple when anxious and pant-grunting. When using right hand, simultaneously fumbles right nipple with thumb and left nipple with little finger, but when he using left hand, fumbles left nipple with thumb and right nipple with little finger. Nipples are massaged by circular movement of thumb and finger. Captive adolescent male bonobo did same in self-reassurance (de Waal and Lanting 1997). (Type A) Category 1–8. See video.

fumble penis (FMP)

Male occasionally fumbles his penis by hand, leading to penile erection. At Mahale, never culminates into ejaculation. For bonobos of Wamba, only one case of fumbling penis, which did not lead to ejaculation (Kano 1998). (Type A) Category 1–7. See video.

fumble penis with foot (FPF)

Rub penis up and down with own heel, toe or sole. Few juvenile, adolescent and adult males do so, without ejaculation (Nishida et al. 2009). Cf. <shake penis>. (Type A) Category 8. See video.

funny face (FUF)

Adult male, Alofu, shows unusual facial expression to estrous females and adult males. Function unknown, but as he showed funny face to female unwilling to follow him despite frequent solicitation for consortship, function may be to reassure. (Type A) Category 9. See video.

G

gallop (GLP)

Goodall (1989): “Fastest run of the chimps.” van Hooff’s (1973) Gallop. Pika’s (2007) Gallop? (Type A) Category 2. See video.

gang attack

See <attack concertedly>.

Gashaka-Gumti

Study area of *Pan troglodytes vellerosus* within Gashaka-Gumti National Park, Nigeria (7°19′N, 11°35′E). Research organised by V. Sommer from 2002 to present (Fowler and Sommer 2007).

gaze (GAZ)

Includes <stare fixedly>, <peer>, and <watch>. (Type D) Category 1. See video.

genital inspection

See <inspect genitals>.

genito-genital contact

See <rub genitals>.

geophagy

See <eat termite soil>.

gestation period

Average = 229.4 days, range = 203–244 days (Martin et al. 1978).

G-G rub

See <rub genitals>.

give (GIV)

Possessor hands over food to another who may or may not have begged for it. Food given includes meat, fruit, leafy branch, and dead wood. Size of given food not necessarily large (Zamma 2005). Rare at Mahale. Unknown in bonobos of Wamba (Kano 1998). (Type C) Category 1–8. See video.

T. Nishida et al., *Chimpanzee Behavior in the Wild: An Audio-Visual Encyclopedia*,
DOI 10.1007/978-4-431-53895-0_8, © Springer 2010

give birth (GBI)

Childbirth usually occurs in bed at night. See also <travel alone after childbirth>. (Type D) Category 1.

glance (GLN)

Goodall (1989): "…looks toward another for less than 2 s." Subordinate chimpanzee looks only briefly at food, estrous female, or other object of competition when dominant male present. Cf. <look>, <peer>, <stare fixedly>, and <watch>. Kano's (1998) Glance for bonobos. (Type A) Category 1. See video.

glottal cramps

See <choke in tantrum>.

glove (GLV)

Wear cylindrical bark ring on lower arm. Female infant, Imani, did this once. (Type A) Category 9. See video.

gnaw

See <mouth>.

go ahead (GOA)

Travel ahead of party, e.g. mother goes ahead of infant. Cf. <depart together>. (Type C) Category 1. See video.

Gombe

Study area of *Pan troglodytes schweinfurthii* in Gombe National Park, Tanzania (4°40′S, 29°38′E). Research organized by Jane Goodall, continuing from 1960 to present (Goodall 1986; Pusey et al. 2008).

Goualougo

Goualougo Triangle, study area of *Pan troglodytes troglodytes* in Nouabale-Ndoki National Park, Republic of Congo (2°05′N–16°56′N, 3°03′E–16°51′E). Research organized by C. Sanz and D. Morgan from 1999 to present (Sanz et al. 2004).

grab (GRB)

Goodall (1989): "…one individual roughly seizes another with one or both hands. This is an aggressive gesture. It is usually followed by a sequence of aggressive interactions, …and may escalate into fight." <Grab> occurs in aggression, grooming and play. Bonobos grab in same contexts (Kano 1998). (Type A) Category 1. See video.

grab and shake (GBS)

Grab another's body part (e.g. leg, foot) and repeatedly shake it. Threatening but also playful. Also, as solicitation of play. (Type A) Category 3–7. See video.

grapple (GPL)

Goodall (1989): "When two individuals go into a clinch with arms locked and heads bowed, and sometimes roll over and over. Individuals may be fighting or playing. Not uncommon when adult females fight one another." Bonobos do same (Kano 1998). (Type A) Category 1. See video.

grasp (GRA)
Grasp object such as branch to carry it or to support body. Infant may grasp mother's nipple during rest. (Type A) Category 1. See video.

grasp and heave (GHV)
Grasp partner's shoulders with both hands, fall on back (supine), kick upwards to partner's belly, pushing partner forward and away. (Type B) Category 1–7. See video.

grasp and push shoulders (GPS)
Infant, Mitsue, in bipedal posture, grasps hair of mother's shoulders and pushes them upward, so mother departs with Mitsue on her back. Soliciting departure likely originated as intention movement of dorsal travel. (Type A) Category 8–9. See video.

grasp hand (GSH)
Goodall (1989) Hold hand. Dominant individual grasps hand extended by subordinate for reassurance, but usually without shaking motion. Also in <hand-wrestle> play. Absent in bonobos of Wamba (Kano unpublished). (Type A) Category 3–5. See video.

greet (GRE)
Friendly reunion as Goodall's (1989) Greeting behavior: "Typical friendly behavior includes <bob>, <crouch>, <touch>, <kiss>, <embrace>, <groom>, <present>, <mount>, <inspect genitals>, <grasp hand>, etc. Vocalizations associated with greeting behavior are <pant-grunt>, <pant-hoot>, <grunt>, <pant-bark>, <scream>, etc." In bonobos, genital contact behavior and <peer> may be comparable to greeting, in addition to <groom>, <present>, <mount>, <inspect>, and <grasp hand>, but the other greeting elements listed above are lacking (Kano 1998). (Type D) Category 1–2. See video.

grimace
See <grin>.

grin (GIN)
Goodall (1989): Facial expression with "the corners of lips are drawn back, exposing the lower, or both upper and lower teeth (or teeth and gums)". Includes van Hooff's (1972) Silent bared-teeth display. Often shown by chimpanzee threatened by another, when copulating (female), begging for foods, or simply watching dominant chimpanzee. May accompany male courtship. Facial expression sometimes accompanies vocalizations such as <scream> and <squeal in copulation>, and may reflect fear or frustration; but also reflect pleasure during copulation (de Waal 1988). An adult male, Fanana, occasionally shows grin and "food screams" during feeding on fruits such as lemon or *Saba* fruit or termites, which seems to express pleasure. (Type B) Category 2. See video.

grin-full-closed (GFC)

Goodall (1989) Full closed grin: “Both upper and lower teeth (and gums) are revealed by the horizontal retraction of upper and lower lips, and the teeth are closed, or almost closed”. Plooij’s (1984) Grin-full-closed (GFC). Kano’s (1998) Grin for bonobos of Wamba. (Type A) Category 2. See video.

grin-full-open (GFO)

Goodall (1989) Full open grin: “Both upper and lower teeth (and gums) are revealed by horizontal retraction of upper and lower lips, and the teeth are partially or widely open.” Plooij’s (1984) Grin-full-open (GFO). Grin for bonobos at Wamba (Kano 1998). (Type A) Category 2. See video.

grin-low-closed (GLC)

Goodall (1989) Low closed grin: “Only the lower teeth (sometimes with gums) are revealed, and the teeth are closed, or almost closed.” Absent in bonobos? (Type A) Category 3. See video.

grin-low-open (GLO)

Goodall (1989) Low open grin: “Only the lower teeth (sometimes with gums) are revealed, and the teeth are partially or widely open.” Absent in bonobos? (Type A) Category 2–3. See video.

groin pocket

See <transport in groin pocket> or <hold object in groin pocket>.

groom (GRM)

Goodall (1989) Grooming behavior: “…use both hands, pushing the hair back with the thumb or index finger of one hand and holding it back while picking at the exposed skin with the nail of the thumb or index finger of the other. The chimpanzee can also use one hand, parting the hair in the same way and holding it back with the lower lip.” Grooming may occur in bipedal, quadrupedal, sitting, or reclining posture. Grooming functions include appeasement, reassurance, reconciliation, and hygiene. (Type B) Category 1. See video.

groom-branch-clasp (GBC)

Groom mutually like <groom-hand-clasp>, but non-grooming hand is not clasped, but held overhead on branch. Seen at Gombe and Mahale. Whiten et al.’s (1999) Branch-clasp. Also recorded for bonobos of Lomako (Hohmann and Fruth 2003). (Type A) Category 2. See video.

groom briefly (GBR)

Groom another individual (<groom unilaterally>) briefly (<10″) in greeting or appeasement. (Type C) Category 3. See video.

groom by hand (GRH)

Groom another with one or both hands. Seem to groom important companions such as relatives and allies with two hands, but others with one hand and often leisurely. (Type A) Category 1. See video. See also videos “groom with mouth” and “smack lip.”

groom carcass (GRC)

Groom carcass or even skin of animal such as colobus monkey or leopard. Function unknown, but shown by most adult members of M group. See also <care alloparentally for another species>. (Type B) Category 3–7. See video.

groom, dyadic (GRD)

Two chimpanzees, either one or both grooming. <Groom unilaterally>, <groom mutually>, or <groom reciprocally>. (Type B) Category 1. See video.

groom ground

See <groom object or substrate>.

groom-hand-clasp (GHC)

McGrew and Tutin's (1978) Grooming-hand-clasp. Kind of mutual grooming at Mahale, Kibale, Lopé, Budongo, Taï and Semliki (Webster et al. 2009), but not at Gombe or Bossou. In captivity, done by chimpanzees at Yerkes Primate Research Center (de Waal and Seres 1997; Bonnie and de Waal 2006). Seen in bonobos of Wamba (Kano 1998), Lomako (Hohmann and Fruth 2003) and Lui Kotale (Fruth et al. 2006). (Type A) Category 2–5. See video. See also video "groom-hand-clasp unilaterally."

groom-hand-clasp unilaterally (GHU)

One grooming partner assumes typical hand clasp posture, but does not groom companion; instead rests or grooms ground, although being groomed. (Type A) Category 8–9. See video.

groom leaf (GRL)

Goodall (1989) Leaf-grooming: "A leaf may be suddenly picked, seemingly at random. Holding the leaves in one hand, the chimp makes grooming movements with both thumbs, often pushing his lower lip to the leaf surface as in social grooming. Other chimps often crowd around to look. After a few minutes the leaves are discarded. It seems to be a form of displacement or redirection." At Mahale, chimpanzee puts small louse in folded leaf and kills it (Zamma 2002a). Louse egg is also crushed during leaf-grooming (Zamma 2006). Customary at Gombe, Mahale, Kibale and present at Budongo (Assersohn et al. 2004) , but not seen at Taï and Bossou, nor in bonobos of Wamba (Kano 1998). See also <inspect leaf> and <squash leaf>. (Type B) Category 4–5. See video.

groom mutually (GMU)

Goodall (1989) Mutual grooming. Two chimpanzees groom each other simultaneously. Common in chimpanzees (Boesch and Boesch-Achermann 2000), but occurs in Wamba bonobos only during role changes of social grooming (Kano 1998). Also rare at Lomako (Hohmann and Fruth 2003). Humans may not groom mutually. (Type A) Category 2–3. See video.

groom object or substrate (GRO)

Groom soil, leaves on ground, tree trunk, or moss on rock, usually with one hand. Solitary or social grooming, even during grooming-hand-clasp. Seen in only two young adult males of Mahale in 1992, but done also by three adult or adolescent females in 1995. (Type A) Category 8. See video.

groom, polyadic (GRP)
Simultaneously, three or more individuals groom in tandem or mutually. See Nakamura (2003). (Type B) Category 3–6. See video.

groom reciprocally (GRR)
Goodall (1989) Reciprocal grooming. Two chimpanzees alternately groom each other. Most common type of social grooming among adults. Common in chimpanzees (Boesch and Boesch-Achermann 2000) and bonobos (Kano 1998). (Type B) Category 1. See video.

groom self (GRS)
Goodall (1989) Self groom, Plooij's (1984) Self-groom (GRS), van Hooff's (1973) Autogroom. Kano's (1998) Self-groom for bonobos. (Type A) Category 1. See video.

groom unilaterally (GRU)
Goodall (1989) Unilateral grooming. Groom another without reciprocation. Kano's (1998) Unilateral grooming for bonobos. (Type A) Category 1. See video.

groom with mouth (GMO)
Use lips and teeth to remove items when grooming, especially from face. Also in bonobos (Kano unpublished). (Type A) Category 2. See video.

groom wound (GRW)
Groom wound of another individual. Cf. <inspect wound>. (Type A) Category 2–3. See video. See also video "inspect wound".

grope (GPE)
Put arm into tree hole and feel for something to eat. Cf <probe>, <probe with finger>. (Type B) Category 1–3. See video.

ground bed or cushion
See <make ground bed/cushion>.

group (GOP)
Make any association of individuals, temporary or long-term. (Type D) Category 1.

grunt (GNT)
Various utterances including <grunt, aha>, <grunt in bed>, <grunt, extended>, <grunt, food> and <pant-grunt>. (Type B) Category 2.

grunt, aha (GNA)
Food grunts uttered loudly when walking hastily in group to known food patch (Goodall 1986). Probably absent in bonobos of Wamba (Kano unpublished). (Type A) Category 3–5. See video.

grunt, extended (GNE)

Goodall (1989): "A long drawn out sound usually composed of two syllables- ehmmmmm- heard during rest sessions, significance not clear." Mahale chimpanzees often "discovered" because of this sound uttered in response to passers-by or appearance of human beings. (Type A) Category 3–5. See video.

grunt, food (GNF)

Uttered during or just before eating, especially during first minutes. Cf. <aha-grunt>. Bonobo give similar grunt (Kano unpublished). (Type A) Category 2. See video.

grunt in bed (GNB)

Goodall (1989) Nest grunt: "Distinctive double grunt given when an individual begins to look around for a suitable nest site, during nest making, or as he settles down for the night." Same as Boesch's (2009) Night call? Not heard in bonobos of Wamba (Kano unpublished). (Type A) Category 3–5.

H

habituated (HAB)

H

No overt response to stimuli, after many repeated presentations that earlier elicited some reaction. Chimpanzees of M group now act normally in human presence although they fled before habituation succeeded.
(Type D) Category 1. See Fig. 18.

Fig. 18 Habituated: Chimpanzees of M group are followed by researchers without any apparent influence on their behavior (T. Nishida)

hair erection

See <bristle>.

hammer nut with stone (HMS)

Use stone to hammer nut on stone anvil or wood anvil: Customary at Bossou and Taï (Whiten et al. 1999); Ebo (Morgan and Abwe 2006); also at Nimba (Sugiyama 2008). At Bossou, only stone anvil used. At Ebo, stone hammer used in tree. Absent at Mahale. (Type A) See video (Bossou).

hammer nut with wood (HMW)

Use wooden club to hammer nut on wooden or stone anvil. Customary only at Taï (Whiten et al. 1999). Absent at Mahale. (Type A) See Fig. 19.

Fig. 19 Hammer nut with wood: Drawn from photo of chimpanzees of Taï (By courtesy of Michio Nakamura)

hand clap

See <clap hand>.

hand-clasp

See <groom-hand-clasp>.

hand-rubbing

See <rub hand/feet>.

hand support

See <transport with hand support>.

handedness
Preference for use of right or left hand. Handedness for tool use was found on individual basis, but not found population-level among wild chimpanzees (Boesch 1991b; Sugiyama et al. 1993; McGrew and Marchant 1999, McGrew et al. 1999; but see Lonsdorf and Hopkins 2005).

handicap self (HCS)
Older individual inhibits rough acts and shows gentle restraint in social play, especially with youngsters, e.g. <lie supine>, <lie prone>, <somersault>, <flee>, <fall over backward>, etc. (Type D) Category 1. See video.

handwrestle
See <wrestle with fingers>.

hang (HAN)
Suspended freely by one or both hands from tree branch (Cf. <brachiate>). Neither foot touches substrate. Kano's (1998) Suspend in bonobos. (Type A) Category 1. See video.

H

hang and spin (HAP)
Hang from horizontal branch with one hand, and revolve repeatedly. Small infants and (also rarely adolescents) show as solo play. Sometimes, hanging immature plays with, or solicits play from, individual resting on ground. (Type A) Category 1. See video.

hang and stamp (HAS)
Hang by one or both hands from large bough and stamp on ground (or sometimes play partner) in solo or social play. May repeat bouncing up and down like trampoline. Infants, juveniles and adolescents enjoy this. (Type A) Category 3–6. See video.

hang in sloth position (HSP)
Goodall (1989) Sloth position: "Hanging underneath a branch using all hands and feet or a combination of any three. Usually part of locomotor play in infants." At Mahale done as solo play by immature chimpanzees. Rarely locomote in this position (see <walk in sloth position>). Bonobos not only hang, but also travel in this posture. (Type A) Category 2. See video.

hang object
See <suspend>.

hang-stand (HST)
Hang from branch while one or both feet touch substrate below. Temporary posture changes to another posture or locomotion, or sometimes brief resting posture. Also in bonobos (Kano unpublished). (Type A) Category 1–2. See video.

hang tripedal (HAT)
Hang while gripping branches with feet and one hand. Head and trunk more or less erect. Other hand used to eat fruits or young leaves at ends of high

branches, or extended to the playmate below in social play. Also in bonobos. (Type A) Category 2. See video.

hang upside-down by feet (HUF)
Hang by feet from branch. Rare in infant and juvenile social and solo play and play invitation. Kano's (1998) Upside-down suspension for bonobos. (Type A) Category 2. See video.

hang upside-down by hands (HUH)
Grip branch with hands, flex legs and hang upside down. Infants and juveniles in social and solo play and play invitation. (Type A) Category 2–3. See video. See also video "play with object."

hang with legs pitterpat (HLP)
Hang from horizontal branch and shake legs up and down, e.g., in response to play partner on ground who tries to grasp leg of suspended individual, or to solicit play from partner on ground. (Type A) Category 2–6. See video.

hang-wrestle (HAW)
Two arboreal chimpanzees <hang> by one hand, and each tries to <grab>, <push>, or <pull> other with other hand. Making playmate fall seems to be aim of most interactions. (Type D) Category 2–3. See video.

harass (HAR)
Subordinate chimpanzee threatens dominant one. Adang's (1984) Tease. At Mahale older juvenile and adolescent males pester and threaten adult females by <shake branch>, <slap>, <hit>, <throw branch>, <charge>, etc. until females finally (sometimes after a year) pant-grunt to them (Nishida 2003a). Cf. <tease>. Bonobo adolescent males at Wamba pester adult males with threatening gestures, such as bipedal swagger or branch-dragging, and so dominate him. Termed as "adolescent harassment" by Kano (1998). (Type D) Category 2. See video.

head down hip up
See <stand with head down, bottom up>.

head tip
See <tip head>.

headstand (HSE)
Stand on one's head to solicit play and in social play. Done by mature and immature individuals. (Type A) Category 3–7. See video. See also video "circle quadrupedal."

heel kick
See <kick heel>.

heel-tap
See <tap heel>.

herd (HER)

Complex of male behavior to control travel of cycling female, so that she follows him. Includes waiting such as <sit>, <glance>, <wait for companion>, <reverse>, and <look back>, threatening such as <stare fixedly>, <shake branch>, <club>, and <scratch dry leaves>, and aggression such as <hit> and <kick>. Alpha male pulled female during herding when she did not follow him. Thus, herding maintains consortship and possessiveness. Cf. <consort> and <lead>. Bonobos show similar behavior but without aggressive components (<hit>, <kick> or <club>). (Type D) Category 2–3. See video.

herd in coalition (HEC)

Watts (1999) Coalitionary mate guarding: "pairs or trios of top-ranking males engaged in cooperative aggression to prevent estrous females from mating with other males, but tolerated each other's mating activities". May be demography-dependent male mating strategy in unit group that has unusually many adult and adolescent males. Not seen at Mahale except for few hours. (Type D) Category 3–5.

hesitate (HET)

Uncertain to do something, for example, repeatedly advance to and retreat from targeted individual. Stimuli include aggressive posture or facial expression of targeted individual, presence of others nearby, or close proximity of observers or dangerous animal such as warthog. (Type D) Category 1–3. See video.

hiccup (HIC)

Involuntary contraction of diaphragm produces distinctive sound. Sounds like humans. (Type A) Category 1. See video.

hide (HID)

Goodall (1989): "Unhabituated chimps may hide behind a tree trunk, or pull a thickly foliated branch in front of them when a human approaches". Similar behavior seen at Mahale (Nishida unpublished) and Taï (Boesch and Boesch-Achermann 2000). Unhabituated bonobos do same thing (Kano 1998). Cf. <detour>. (Type D) Category 1–2.

hit (HIT)

Bring down an arm from above and strike partner, animal or insect, etc. with fist or knuckles. van Hooff's (1973) and Goodall's (1989) Hit includes striking with palm, here called <slap>. <Hit> either in play or in attack (Goodall 1989). Kano's (1998) Hit and Pika's (2005, 2007) Punch in bonobos. (Type A) Category 1. See video.

hit and run (HIR)

Hit or touch potential playmate and flee to solicit play. (Type B) Category 1–8. See video.

hit bush bipedal (HIB)

Stand bipedal and hit bush with one or both arms, in charging display, threat or soliciting play. (Type A) Category 7. See video.

hit ground with fist (HGF)

Hit ground with fist. Middle-aged mother, Juno, standing on all fours, hit ground with fist, watching her infant nearby. Infant did not respond and Juno grinned. Middle-aged female, Opal, hit ground and pant-grunted when alpha male approached. Juvenile male, Xmas, hit ground with fist as solo play. Adolescent male, Cadmus, did so with fists alternately few meters from large grooming clique. Function probably attention-getting. (Type A) Category 8. See video.

hit toward (HTT)

Threatening gesture. Goodall (1989): "When an animal is threatening it will make a hitting movement with the back of its hand. This gesture is usually accompanied by a soft bark. It may be directed toward chimps, other primates, snakes, large insects, birds, and so forth." Cf. <raise arm quickly> and <shake arm>. (Type A) Category 3–5.

hit with long branch

See <club>.

hold body part in mouth (HBM)

Hold by mouth body part such as fingers, hand, toes, foot, nape, limb or back of companion, usually in play. (Type A) Category 3–5. See video.

hold down

See <press down>.

hold finger(s) in mouth

See <hold body part in mouth>.

hold foot in mouth

See <hold body part in mouth>.

hold genitals (HLG)

Goodall (1989) Hold genitals. Adult grabs or touches scrotum or penis of another as reassurance. Often occurs when adult male mounted by another male; mounted reaches back between his thighs and holds scrotum of mounter. When adult female pant-grunts to adult male, she may hold his scrotum. Absent in bonobos of Wamba (Kano 1998). (Type A) Category 3–5.

hold hand

See <grasp hand>.

hold hand in mouth
See <hold body part in mouth>.

hold head or face (HHF)
Support another's face or head with hand while grooming it with other hand or with lips and teeth. Seen in bonobos (Kano unpublished). (Type A) Category 1–2. See video.

hold object in groin pocket (HLS)
Keep object in fold of thigh and groin. See also <transport in groin pocket>. (Type A) Category 3–5. See video.

hold object in mouth (HLM)
Keep object between jaws. Small infant often keeps twig or leaf in mouth and nips off pieces in solo play. Small stone also held. See also <solicit play with object in mouth> and <transport in mouth>. (Type A) Category 1. See video.

hold object in neck pocket (HNE)
Keep object in fold between lower jaw and flexed upper shoulder. (Type A) Category 3–5. See video.

hold object on head (HLH)
Keep object on head. Adult female, Opal, held her dead infant on head. See <transport on head or nape>. (Type A) Category 1–3. See video.

hold toe(s) in mouth
See <hold body part in mouth>.

hold up chin or head
See <raise other's chin>.

hoo (HOO)
Goodall (1989): "A single syllable soft whimper. A single hoo may be uttered several times in succession. It is the typical sound given by an infant which cannot reach the nipple and wants to reestablish contact with the mother etc." Mahale infants hoo when begging for food, when wanting to return to mother after separation, and when seeking to mate with estrous female. Mothers also hoo when older infant does not return to her, often while playing overhead when she wants to leave. Hoo expresses frustration. Cf. <whimper>. Absent in bonobos? (Type A) Category 3–5. See video.

hook branch (HOK)
Sugiyama (1981, 2008) Branch-hauling, Whiten et al.'s (1999) Branch-hook: Use branch to hook another branch. Present only at Bossou. Absent at Mahale. (Type A) See Fig. 20.

hoot (HOT)
Goodall (1989): "Roar pant hoot. Continuous low-pitched calls given only by highly aroused individuals, rarely by females. Always accompanied by a charging display." (Type A) Category 3–5.

Fig. 20 Hook branch: Chimpanzees of Bossou hooked branch overhead with stick (By courtesy of Yukimaru Sugiyama)

hoot face (HOF)

Goodall (1989) Hoot face: "…pouted trumpeted lips that go with pant-hooting. The mouth may open completely or partially." (Type A) Category 3.

hop bipedal on spot (HBS)

Nishida (1997) Jump bipedal. Juvenile male in courtship jumps in place in front of estrous female, to elicit her to <present>, or in front of adult female to harass her. Some juveniles jump up and down while pant-grunting vigorously in front of dominant male. Cf. <leap bipedal> and <hop quadrupedal on spot>. (Type A) Category 8. See video.

hop quadrupedal on spot (HOP)

Hop repeatedly on four limbs on ground or on mother's back as solo play. Infant may hop while pant-grunting. Courtship by juvenile male. Also hop on branch, vegetation, bed, etc. as trampolining. Cf. <leap quadrupedal>. (Type A) Category 3–5. See video.

hug (HUG)

Caretaker embraces infant with arms. Used for consolation or intention movement of mounting. Kano's (1998) Protective embrace in bonobos. Cf. <rush to embrace>. (Type A) Category 1. See video.

hunch (HUC)

Goodall (1989): Hunching shoulders in "contexts of aggression, courtship and greeting." "Nearly always accompanied by hair erection." Plooij's (1984) Hunch (HHC). (Type B) Category 2–3.

hunch and sit (HUS)

Goodall (1989) Sitting hunch: "…while sitting, raises shoulders, arms held out from the body either to the side or in front." At Mahale, sitting hunch courtship often accompanied by stamping ground or shaking branch. Bonobo male's "Sitting erect" (Kano 1998) may be homologous. (Type A) Category 2–3. See video.

hunch bipedal (HUB)
Goodall (1989): "Chimp stands on one's feet, shoulders hunched up, arms held slightly out and away from the body. If he sways from foot to foot, known as bipedal swagger". Also in Kanka Sili (Albrecht and Dunnett 1971). Cf. <swagger bipedal>. Absent in bonobos. (Type A) Category 3. See video.

hunch over (HUO)
Goodall (1989): "Similar to bipedal hunch, but arms held forward over the back of another chimp." de Waal's (1988) Hunch over and Kano's (1998) Bluff over for bonobos. (Type A) Category 2. See video.

hunch quadrupedal (HUQ)
Goodall (1989): "Stand, walk or run on all fours with rounded back and head pulled in between shoulders." Seen in adult males at Mahale. Also in bonobos (Kano unpublished). (Type A) Category 2. See video.

hunt (HUN)
Behavioral complex of stalk, pursue, capture, and kill large birds (e.g. guinea fowl) or small to medium-sized mammals. Include <monitor>, <climb>, <stalk>, <run>, <chase>, <leap on>, <grab>, <flail>, <drag>, <drag to kill>, <throw>, <bite>, <pull>, etc. Known throughout chimpanzee range (Uehara 1997; Mitani et al. 2002). Bonobos hunt less often than chimpanzees (Kano 1992, but see also Surbeck and Hohmann 2008). See also <eat meat>. (Type D) Category 1. See video.

hunt with tool (HWT)
Thrust stick to expel or to skewer prey from tree hole or rocky cavity at Mahale and Fongoli. Prey species are squirrel, hyrax, and bushbaby (Huffman and Kalunde 1993; Pruetz and Bertolani 2007; Nakamura and Itoh 2008). See also <expel>. (Type B) Category 3–6. See video.

hurl self (HUR)
Throw oneself at another individual. Play pattern employed in tree. (Type B) Category 2–8. See video.

huu (HUU)
Goodall (1989): "...very like the hoo whimper, but is usually higher pitched and the huu does not show the pouted lips characteristic of hoo. It is made when a chimpanzee suddenly hears or sees a strange object or sound." Sound often helps researcher to find otherwise silent, hidden chimpanzee in bush. Absent in bonobos? (Type A) Category 3–5.

I

ignore (IGN)

Adult female does not respond to harassment by immature male, so avoids escalation of harassment. Mother does not approach weanling despite its request to suckle, ride, etc. Adult male does not respond to other's pant-grunt. Goodall's (1989) Ignore ("When an individual, sometimes dominant, does not respond to the gesture in any manner apparent to the human observer.") includes both <ignore> and <snub>. Kano's (1998) Ignore for bonobos. Cf. <snub>. (Type C) Category 1. See videos.

imaginary play

See <play, imaginary>.

imitate (IMI)

Goodall (1989) Imitation "When one chimp, after watching the behavior of another, then does the same." (Type D) Category 1. See video.

immature

Age-class consisting of infancy, juvenility and adolescence (age <13 years for females and <16 years for males).

immerse

See <dunk>.

immigrate (IMM)

Adolescent female leaves natal group to enter neighboring group for reproduction. Immigrant usually welcomed by males, but threatened by adult and adolescent females of new group. Juveniles inspect immigrant. See <transfer>. (Type D) Category 1. See video.

immobilize (IMB)

Two or more attackers restrain single mature chimpanzee, using <hold>. Tactic used in lethal coalitionary attack on victim. Behavior includes <hold>, <pull>, <bite>. Recorded or suspected to occur at Gombe (Goodall 1986), Mahale (Nishida et al. 1995), and Kanyawara (Muller 2002). Cf. <press down>. (Type D) Category 3–5.

incest (INC)

Copulate with close relatives. Mother and infant son copulate during weaning period at Mahale. Mating between mother and adult son once observed. Mother and adolescent son or brother and sister mating not seen at Mahale, although juvenile male once tried to mate with mother but was refused. At Gombe, copulation between mother and mature son, and brother and sister rarely seen (Goodall 1986) and alpha male produced offspring with mother (Constable et al. 2001). See <avoid incest>. (Type D) Category 3–5. See video.

index-hit

See <squash ectoparasite on arm>.

infant

Youngster from birth to weaning. 0–4 years old. See Fig. 21.

Fig. 21 Infant: Three-year-old female (T. Nishida)

infanticide
See <kill infant> and <eat infant>.

infantile bark
See <bark, infantile>.

insect-pound
See <pound insect>.

inspect (INS)
Investigate object (including conspecific) by <peer>, <gaze>, <sniff>, <touch>, <climb>, etc. Cf. <monitor>. (Type D) Category 1. See video.

inspect fruit (INF)
Travel through concentrated patch of fruit trees, apparently to check quality of crop, even long before fruit ripens. Includes <walk>, <watch>, <touch fruit>, <sniff>, and <bite>. (Type D) Category 1–5. See video.

inspect genitals (ING)
Goodall (1989) Inspect: "....male or female touches the vaginal opening of a female and sniffs finger, or sniffs with nose directly. Both hands used to pull apart the lips of vagina." Plooij's (1984) INS. Nishida's (1997) Genital inspection. Recoded from Taï (Boesch 2009). Adult male bonobos do not inspect female genitalia, but juveniles do (Kano 1998). (Type B) Category 2–3. See video. See also video in "clip leaf by mouth."

inspect leaf (INL)
Whiten et al. (1999) Leaf-inspect: Inspect ectoparasite on leaf. Component of leaf-grooming. During social and self grooming, take leaf, bring to lips, fold leaf with finger and thumb, open leaf and often inspect something on leaf. Twice confirmed to be louse or its egg (Zamma 2002a, 2006). Customary at Budongo and Mahale and present at Gombe. Pattern at Budongo may be cultural variant (Assersohn et al. 2004). (Type B) Category 5.

inspect self (INA)
Inspect own body-part, e.g. genitals. May be combined with <groom self>. (Type D) Category 1–3. See video.

inspect wound (INW)
Show keen interest in own or other's wound. May approach another to <peer>, <touch>, <lick>, <mouth>, or <groom> wound. (Type B) Category 3. See video.

interact with human (IWH)
After habituation, some chimpanzees of Mahale began to communicate with humans by <extend hand>, <touch>, <pull>, <slap>, <push>, <throw stone or rock>, <throw branch>, <club ground>, etc. Some agonistic and others playful. Cf. <play with another animal>. (Type B) Category 8.

interfere (ITF)

Prevent another from engaging in self-maintenance or social behavior. Female weanling, Aqua, interfered with consort grooming her mother. Interference in dominance interactions is <intervene>. Bonobos interfere in sexual and grooming interactions (Kano 1998). (Type D) Category 1. See video.

interfere fishing (ITH)

Infant often reaches hand to hand or fishing probe of mother or older siblings who is fishing for ants, trying to snatch probe or ants. This act much interferes with performer's activity. Mother often allows infant to take probe. (Type D) Category 3–7. See video.

interfere in copulation (ITC)

Weaning infant, juvenile or adolescent male, often offspring of mating female, runs to push itself between mating pair, waves arms, or touches or pushes at copulating male, while emitting whimper-scream or loud scream (Tutin 1979b). Juvenile male may interfere with elder brother's mating. Dominant adult male may interfere in others' mating by bark, threat, or direct charge. Adult female (often in estrus) may squeeze herself between mating pair (Nishida 1997). Dominant adult male bonobos interfere in mating of adult females, which Kano (1998) called Sexual interference. (Type D) Category 2. See video.

interfere play (ITP)

Third party prevents ongoing social play by rushing at player, sometimes stopping play. Mother may prohibit rough alloparent from continuing play with her infant. Intruder may solicit play in various ways, 'take over' play partner and engage in social play. (Type D) Category 1–3. See video. See also video "kick".

inter-group behavior (IGB)

Behavior related to presence or absence of neighboring unit groups. Generally, antagonistic. Groups usually avoid one another, but larger party of unit group occasionally intrudes territory of another group and kills some of them, in particular infants. However, young females in estrus are allowed to enter neighboring group. See <immigrate>, <intrude>, <kill adult male>, <kill infant>, <patrol>, <respond to neighboring unit group> and <transfer>. (Type D) Category 3.

intervene (ITV)

Exert influence on agonistic interaction such as quarrel between two or more others by approach, threat or aggression. Goodall's (1989) Interference: Intervention in which "a third individual prevents or tries to prevent an interaction between two others." Intervener may be partial or impartial. Also in bonobos. Includes <interfere> and <support>. (Type D) Category 1. See video.

intervene to separate (ITS)

Alpha male charges at rival male and third male who are grooming each other or sitting together, apparently to prevent formation or development of coalition between them. de Waal's (1982) Separating intervention. See also Nishida and Hosaka (1996). Absent in bonobos, but dominant individual (female more often than male) thrust torso, or arm between grooming pair to take over grooming (Kano unpublished). (Type D) Category 3. See video.

intimidate

See <threaten>.

intrude (INT)

Wrangham (1999) Deep incursion. Patrol crosses group boundary and invades territory of neighboring group to attack its members or to explore for resources such as food or mates. Recorded at Gombe (Goodall et al. 1979), Mahale (Nishida 1979; Itoh et al. 1999), Kanyawara (Wrangham 1999), Ngogo (Watts et al. 2006), and Taï (Boesch and Boesch-Achermann 2000). (Type D) Category 1–3.

investigation probe

See <probe>.

investigatory probe

See <probe>.

invitation slap

See <slap in invitation>.

invitation stamp

See <stamp in invitation>.

invite

See <solicit>.

J

J

join play (JOI)

Small infant rushes into vigorous play between seniors and interferes with them by <approach>, <touch>, <slap>, <hit>, <kick>, <wedge>, <hug>, <mouth>, etc. Infant rarely allowed to join. Adult, mother, adolescent or juvenile also joins play group of immatures and often succeeds in playing with one or two of them. (Type D) Category 3–6. See video.

jump

See <leap> and <hop>.

juvenile

Weaned youngster follows mother in daily ranging most of time. 5–8 years old. See Fig. 22.

Fig. 22 Juvenile: Eight-year-old male (T. Nishida)

K

Kabogo

Study area of *Pan troglodytes schweinfurthii* in western Tanzania (5°28′S, 29°45′E). Research done from 1961 to 1963 by K. Imanishi and colleagues (Azuma and Toyoshima 1963).

Kahuzi

Study area of *Pan troglodytes schweinfurthii* in Kahuzi-Biega National Park, Democratic Republic of Congo (2°S, 28°E). Research organized by J. Yamagiwa from 1991 to present (Yamagiwa et al. 1996; Basabose and Yamagiwa 2002).

Kalinzu

Study area of *Pan troglodytes schweinfurthii* in Kalinzu Forest Reserve, Uganda (0°17′S, 30°07′E). Research organized by C. Hashimoto and T. Furuichi, continuing from 1995 to present (Hashimoto and Furuichi 2006).

Kanka Sili

Study area of *Pan troglodytes verus* near Kindia (10°03′N, 12°52′W) in Guinea. Pioneer studies done by A. Kortlandt (Kortlandt and Kooij 1963) in the early 1960s and his colleagues (Albrecht and Dunnett 1971) in 1968–1969.

Kanyawara

Study area of *Pan troglodytes schweinfurthii* in Kibale National Park, Uganda (0°34′N, 30°21′E). Research organized by R. Wrangham, continuing from 1987 to present (Wrangham et al. 1996).

Kasakati

Study area of *Pan troglodytes schweinfurthii* in western Tanzania (5°27′S, 29°55′E). Research done from 1963 to 1967 by J. Itani, K. Izawa, M. Kawabe, and A. Suzuki (e.g. Izawa 1970, Suzuki 1969).

Kasoje

Northwestern part of Mahale Mountains National Park, with headquarters of research and ecotourism.

keep water in mouth (KWM)

Drink from stream and sometimes retain water in mouth for more than 10 s before swallowing. Function unknown. (Type A) Category 1–6. See video.

kick (KCK)

Goodall (1989): "Make contact with an objective (usually another chimp) with one or both feet. Kicking is a forward, sideways or backward movement, different from a stamp which is always downward." In aggression (Nishida 1994) and play. See also <drum>. Cf. <stamp>. Many types of kicking described for bonobos by Kano (1998). (Type B) Category 1. See video.

kick back (KCB)

Goodall (1989): "Walking individual kicks backwards towards a following youngster…" Seen in bonobos (Kano 1998). (Type A) Category 2. See video.

kick backward quadrupedal (KCW)

Kick backward to buttress, tree trunk, wall, etc. while standing on four limbs. Goodall's (1989) Drum/Kick. (Type A) Category 3–6. See video.

kick bipedal (KCP)

Kick buttress or tree trunk with foot while standing bipedal. See also <drum>. Absent in bonobos? (Type A) Category 3–5. See video.

kick buttress (KKB)

Kick buttress or alternative such as tree trunk and metal wall. Done mainly by adult males. Demonstrate own geographic position. See also <drum>. (Type B) Category 3–5. See video.

kick heel (KCH)

Goodall (1989) Heel kicking: "…resting individual (often an adolescent or adult) raises one leg and thumps its heel down on a youngster who is playing around him or her." Also at Mahale. Goodall's other type is "An older male, when lying on back holding an infant in a ventro-ventral embrace, kicks down gently one heel after the other, on the infant's rump." Absent in bonobos (Kano 1998). (Type A) Category 3–5. See video.

kick other (KCS)

Kick another in fight and play. Seen in bonobos (Kano 1998). Pika's (2007) Kick. (Type C) Category 1. See video.

kick up (KCU)

Lie supine and kick into air object, such as carcass. If carcass, may be conflict behavior. (Type A) Category 3–6. See video.

kidnap (KDN)

Nishida (1983a) Detach: Take infant from mother and keep it under alloparent's control for as long as alloparent wishes. Hard for infant to return to mother if taken far away from her. Goodall's (1989) Kidnap: "Borrow by consent. When an infant is carried away from the mother by a sibling or by another chimp." See also <care alloparentally>. Seen in bonobos (Kano 1998). (Type D) Category 1. See video.

kill (KIL)
Fatal attack to another. (Type D) Category 1–3.

kill adolescent male (KAM)
At Gombe, two cases of between-group fatal attack of adolescent males seen or suspected (Wilson et al. 2004). (Type D)

kill adult female (KMF)
Adult females killed within group at Ngogo (Watts 2004) and between groups at Gombe (Goodall 1986). Adult female attacked between groups but rescued by researchers at Mahale (Nishida and Hiraiwa-Hasegawa 1985). (Type D) Category 1–5.

kill adult male (KMM)
At Mahale, ex-alpha male probably killed within group (Nishida 1996) and several adult males between groups (Nishida et al. 1985). One case of within-group (severe but non-fatal) attack on young adult male, considered to be "ostracism" (Nishida et al. 1995). Killing adult male within group recorded or suspected at Gombe (Goodall 1992), Budongo (Fawcett and Muhumuza 2000) and Ngogo (Watts 2004) and killing of adult males between groups at Gombe (Goodall 1986), Kanyawara (Wrangham 1999; Muller 2002), Ngogo (Watts et al. 2006) and Kalinzu (Hashimoto and Furuichi 2005). (Type D) Category 1–5. See Fig. 23.

Fig. 23 Kill adult male: Ex-alpha male, Ntologi, suspected to have been killed by group members, given so many wounds on body (T. Nishida)

kill another species (KAN)

Kill another species for eating or teasing by <flail>, <knock down>, <throw>, <drag to kill>, <hit>, <bite>, etc. (Type D) Category 1. See video.

kill infant (KIF)

Infanticide. Infants (at least 8) killed at Mahale, with most eaten by killers and others: between-group 3 times (Nishida et al. 1979; Kutsukake and Matsusaka 2002) and within group >5 times (Hamai et al. 1992). Between-group infanticide recorded from Budongo (Newton-Fisher 1999), Gombe (Goodall 1986), and Ngogo (Watts et al. 2002). Within-group infanticide recorded at Gombe (Goodall 1977, 1986) and Kanyawara (Arcadi and Wrangham 1999), and suspected at Taï (Boesch and Boesch-Achermann 2000). Most infanticide was carried out by adult males, but occasionally by adult females (Goodall 1977; Townsend et al. 2007). No infanticide in bonobos of Wamba (Kano unpublished). See also <eat conspecific>. (Type D) Category 3. See Fig. 24.

Fig. 24 Kill infant: Alpha male, Ntologi, killed and ate infant (By courtesy of Tamotsu Asou)

kill juvenile (KIJ)

Between-group fatal attack on juvenile male reported from Ngogo (Watts et al. 2006) and suspected at Gombe (Williams et al. 2002). Absent at Mahale. (Type D)

kill time (KLT)

Behavior when waiting for long time, e.g. when close companion sleeps for long bout, sits, grooms, fishes for ants or termites, etc. Includes <yawn>, <flip lip>, <drop bark bits>, <scratch self>, <groom object or substrate>, etc. (Type D) Category 1–3. See video.

kiss (KIS)
Goodall (1989) divides Kiss into two forms: pout-kiss and open-mouth kiss. Both types rarely seen in bonobos, but absent in adult bonobos at Wamba (Kano 1998). (Type B) Category 1. See video.

kiss with open mouth (KOM)
Plooij (1984) OMK. Goodall (1989) Open mouth kiss: "The open mouth is pressed to body or mouth of another. If two chimps open-mouth kiss each other on the face one often presses his mouth over the other's upper lip and nose, the other has his mouth over the lower lip and chin of the first…" Open-mouth kiss often occurs in reunion, reconciliation, or social excitement, especially, among adult males, accompanied by panting. Kano's (1998) Open mouth kiss directed only to infants and juveniles by older bonobos, and not between adults. (Type A) Category 2–3. See video.

kiss with pout face (KPO)
Goodall (1989) Pout-kiss: "The lips are slightly pouted and pressed against, or briefly laid against, the body, face, or limbs of another. Pout kissing often occurs in greeting, submissive and reassurance contexts." Mother kisses infant in head or face for self-reassurance after agitated interaction with other. Even colobus carcass kissed. Absent in bonobos (Kano unpublished). (Type A) Category 3. See video.

kiss with tongue (KTO)
Tongues pressed each other during kissing. Juvenile female and male, Imani and Oscar, showed this "French kiss" for 7 s as reconciliation after quarrel. (Type A) Category 8–9. See video.

knock with both arms (KNB)
Hold, lift, and knock object such as colobus monkey (often even carcass) against substrate (usually ground) with both arms. Kills prey and simultaneously intimidates followers of meat-holder. Also seen at Taï (Boesch 2009). (Type A) Category 3–8. See video.

knock with one arm (KNO)
Hold, lift, and knock object against substrate (usually ground) with one arm. Adolescent male, Cadmus, knocked colobus carcass with one hand, holding branch with other. Kills prey and simultaneously intimidates followers of meat-holder. Also seen at Taï (Boesch 2009). (Type A) Category 3–8. See video. See also video "knock down with both arms."

knuckle-knock
See <rap>.

knuckle walk
See <walk quadrupedal on knuckles>.

L

laugh

See <play-pant>.

lead (LED)

Mature male or female, and especially adolescent male, leads sexual partner into undergrowth or higher in tree to avoid interference by more dominant males (Nishida 1997). See also <consort>. Cf. <herd>. Seen in bonobos (Kano unpublished). (Type D) Category 2. See video.

L

leaf-clip

See <clip leaf>.

leaf cushion

See <make leaf cushion>.

leaf-dab

See <dab leaf>.

leaf-fold

See <fold leaf>.

leaf-groom

See <groom leaf>.

leaf-inspect

See <inspect leaf>.

leaf-midrib spoon (LMS)

Juvenile male, Xmas, and infant female, Liz, removed leaf-blade by clipping leaf and used midrib for licking water. Rare solo play. Cf. <leaf-spoon>. (Type B) Category 8. See video.

leaf-mop

See <mop ants with leaves>.

leaf-napkin

See <wipe with detached object>.

leaf-pile pull
See <pull leaf-pile>.

leaf-pile push
See <push leaf-pile>.

leaf-sponge (LSP)
Remove leaves from tree or shrub, dip into water, then put into mouth, and compress. Result is kind of absorbant sponge, which is soaked into tree-hole or stream, then sucked to drink water (Matsusaka et al. 2006). Whiten et al. (1999)'s Leaf-sponge. Chimpanzees of Tongo used moss for leaf sponges (Lanjouw 2002). Customary at Gombe, Mahale, Kibale, Budongo, Bossou, Taï, and Goualougo (Sanz and Morgan 2007). Young bonobos of Lomako also use moss sponges (Hohmann and Fruth 2003). (Type B) Category 2–7. See video.

leaf-spoon (LSN)
Put leaf into stream and suck, pour or lick water from it (Matsusaka et al. 2006). Cf. <leaf-midrib spoon>. (Type B) Category 8. See video.

leaf-squash
See <squash leaf>.

leaf strip
See <strip lea>.

leaf-swallow
See <swallow leaf>.

lean (LEA)
Incline torso to rest against tree trunk, bough, rock, or another. Kano's (1998) Reclining sit. (Type A) Category 1. See video.

lean forward (LEF)
Flex trunk to expose back for grooming. Seen in bonobos (Kano unpublished). (Type A) Category 2.

leap (LEP)
Jump propelled by feet. Plooij's (1984) JUM. Kano's (1998) Leap and Pika's (2007) Jump in bonobos. (Type B) Category 1.

leap between trees (LET)
Leap horizontally from tree to tree. Done by youngsters rather than adults, during play and when many food trees grow close together. (Type A) Category 2. See video.

leap between trees with object (LEO)
Leap horizontally from tree to tree, while carrying object on shoulders. (Type B) Category 9. See video.

leap bipedal (LPB)

Adult male (occasionally, female) stands upright, swings body forward and back, to gain momentum, then jumps from rock to rock in upright posture when crossing stream on stepping stones. Absent in bonobos of Wamba (Kano unpublished). (Type A) Category 3–6.

leap bipedal with squared shoulders (LSS)

Adult male (or female) with arms half-raised jumps bipedally up and down repeatedly facing another male (or female) doing same. Threatening gesture in adult males and females. (Type A) Category 3–7. See video.

leap down (LPD)

Goodall (1989) Vertical leap: "Leap vertically from one structure to another." Leap from tree down to vegetation or ground during play, at start of departure, or when chased. Infant leaps down from mother's back to ground. Cf. <fall> and <drop self>. Kano's (1998) Vertical leap for bonobos. (Type A) Category 2. See video.

leap down and wait (LPW)

As mother travels, older infant leaps down from mother's back, runs ahead, climbs shrub or fallen log, waits for her arrival, then leaps on her back, and resumes ride. Infant play, repeated again and again. (Type B) Category 3–7.

leap into lap (LAP)

Goodall (1989): "An infant jumps into the lap of a sitting chimp, during greeting." Seen at Mahale, but not in greeting. (Type A) Category 3–6.

leap on (LPO)

Spring onto another in play and aggression. Infant may leap from mother's back to older playmate's back in traveling. Kano's (1998) Jump on similar for bonobos, but jumper likely to mount. (Type A) Category 2. See video.

leap quadrupedal (LPQ)

Usual way of crossing stream via stepping stones. Before leaping, chimpanzees swing body back and forth to gain momentum in quadrupedal posture. Cf. <hop quadrupedal on spot>. Bonobos commonly leap on all fours arboreally and on ground (Kano 1998). (Type A) Category 2. See video.

leap up (LPU)

Jump upward suddenly from sitting to grasp overhead branch or hanging playmate. Done by infant without anything overhead as solo play. Cf. <leap up in surprise>. (Type A) Category 3–7. See video.

leap up in surprise (LPS)

While standing bipedal, may leap up when surprised. (Type A) Category 3–8. See video.

leave (LVE)

Travel away from another with no parting signals. Plooij's (1984) LVE. Cf. <depart>. (Type C) Category 1. See video.

leave and climb (LVC)
Infant leaves mother’s back and climbs tree. (Type B) Category 2–3. See video.

leave to protest (LVP)
Weanling protests to mother by leaving and keeping distance from her, usually with whimper or whimper-scream. See also <monitor mother>. Cf. <fall over backward>. (Type D) Category 3. See video.

leg cradle
See <cradle with leg>.

leopard
See <respond to leopard>.

lever open (LVO)
Whiten et al. (1999) Lever open: Use stick to enlarge entrance. Customary at Gombe, Taï, Lopé, and Goualougo (Sanz and Morgan 2007). Absent at Mahale. (Type A)

lick (LIK)
Plooij (1984) LIK. Repeatedly stroke with tongue. (Type A) Category 1. See video.

lick lips (LLP)
Lick away items on lips. (Type A) Category 1. See video.

lick rock (LIR)
Lick rocks at few sites along shore of Lake Tanganyika or in riverbed of large rivers. More often in dry than wet season (Nishida 1980a). May continue an hour. At Mahale, seen in chimpanzees of B, K, M and L groups (unpublished data). Nutritional significance unknown. Rocks licked not salty to human taste. Not seen at Gombe (Goodall 1986). (Type B) Category 6. See video.

lick snot
See <eat nasal mucus>.

lick wood (LIU)
Licks de-barked surface of dead trees such as *Pycnanthus angolensis*, *Ficus capensis*, *Ficus vallis-choudae*, *Garcinia huillensis*, etc., with extended upward laps of tongue. Surface exudes some sugar alcohol, which might be nutritionally meaningful. Not seen at Gombe. (Type B) Category 6. See video.

lick wound (LIW)
Lick wound apparently to clean it, directly with tongue or touch wound and then lick fingers. Sometimes done to others, often by infants or juveniles. Reported from Gombe, Mahale and Taï. Seen in bonobos, who also lick menstrual blood (Kano 1998). (Type B) Category 1. See video.

lie (LIE)
Recline on ground or other horizontal substrate such as stout branch. Plooij's (1984) LIE. Kano's (1998) Lie for bonobos. (Type B) Category 1. See video.

lie and hug (LIH)
Nishida (1983a): Caretaker lies supine on ground while embracing infant to chest. Seen in bonobos (Kano unpublished). (Type A) Category 1. See video.

lie and watch (LWT)
Adult or adolescent lie on ground and watch playing juveniles or infants to solicit play. (Type B) Category 3–7. See video.

lie lateral (LID)
Lie on side. Common resting or sleeping posture on ground. (Type A) Category 1. See video.

lie on other (LIO)
Lie on other's belly, etc. in play. Larger individual (e.g. adolescent) may lie supine on smaller play partner (e.g. infant), immobilizing it during wrestling. Cf. <step on> in which smaller playmate 'rides' on older reclining companion. Cf. <press down> in which larger individual stands quadrupedal, sits, or lies on stomach. (Type A) Category 3–7. See video.

lie prone (LIP)
Lie on belly with legs and arms flexed. Assumed occasionally when waiting for play partner to come after solicitation of play. Also posture assumed by estrous or pregnant females near term. (Type A) Category 1. See video.

lie-sit (LSI)
Typical sitting or reclining posture of estrous female, apparently forced by very swollen sexual skin. (Type A) Category 3. See videos.

lie supine (LIS)
Lie on back with legs not spread. Most common sleeping posture in bed or on ground. Also, as solicitation of play, in which playmate may <step on> him/her. (Type A) Category 1. See video.

lie supine and shake arms and legs (LSL)
Lie supine and shake arms and legs quickly and irregularly. Solo play of adolescent males, Cadmus and Primus, juvenile male, Xmas, and infant female, Xantip. Occasionally used as solicitation of play, as juvenile male, Michio, did so on back of adult male. (Type A) Category 8. See video.

lie supine with legs apart (LIL)
Lie on back with legs spread. Often occurs when male solicits another to groom scrotum. (Type A) Category 3. See video.

lie with back to another (LIB)
Pattern of soliciting grooming of back. See also <solicit grooming>. Seen in bonobos (Kano unpublished). (Type C) Category 2. See video.

lie with legs crossed (LIX)
Lie supine with raised legs crossed. Resting posture. (Type A) Category 1–3. See video.

lift (LFT)
Raise up heavy object by hands. Seen in bonobos. (Type A) Category 1. See video.

lift and drop (LFD)
Raise up heavy log or thick woody vine by hand(s), then drop it on ground to make loud, thumping sound. Display by young adolescent males, Primus and Orion, and adult male, Pim, as courtship, mild threat, or solicitation of play. (Type B) Category 8. See video.

lift rock (LFR)
Adult male lifts rock with hands, then usually throws into water to make noisy splash, apparently heightening effect of charging display. See <throw splash>. No rocks in habitat of Wamba bonobos. (Type A) Category 7. See video.

Lilungu
Study area of *Pan paniscus* in Equateur Province of DRC (1°7′36″S, 23°31′28″E). Research done by Sabater Pi and colleagues from 1988 to 1990 (Bermejo et al. 1994).

limp (LMP)
Walk lame because of wounded limb. (Type A) Category 1. See video.

lion
See <respond to lion>.

lip flip
See <flip lip>.

lip-smack
See <smack lip>.

listen (LST)
Orient head (and so ears) to source of sound or calls, while stopping on-going activities, such as groomimg or eating. Cf. <search for conspecific>. (Type A) Category 1. See video.

locomote (LCM)
Includes <walk>, <crutch>, <retreat>, <limp>, <stamp trot>, <run>, <gallop>, <somersault>, <pirouette>, <climb>, <leap>, <brachiate>, <sway and move>, <swing and grasp>, <fall>, <descend>, <slide down>, etc. (Type D) Category 1.

locomotion
See <locomote>.

locomotion play
See <play>.

locomotor behavior
See <locomote>.

locomotor play
See <play>.

locomotory aid
See <aid in locomotion>.

Lomako
Study area of *Pan paniscus* in Equateur Province of DRC (0°50′N, 21°05′E). Research initiated by A and N Badrian from 1974 and taken over by many researchers after 1980s (Susman 1984).

lone play
See <play solo>.

longevity
Longest lifespan. About 50 years at Mahale.

look around (LOA)
Look side to side or back and forth, often accompanied by straightening back and tilting head, in order to monitor surroundings. Cf. <scan> and <search for conspecific>. (Type B) Category 1. See video.

look at water (LAW)
Sit or stand quadrupedally and look at water's surface, either to watch object in water, or to view reflection of self-image. Cf. <nod quadrupedal>. See also <play with water>. (Type A) Category 1–8. See video. See also video "drink from stream."

look back (LOB)
Groomee turns to look at groomer who has stopped grooming, and groomer immediately resumes grooming. Thus, turning round is request to resume grooming. Walking chimpanzee looks back to see if companion is following. e.g. when adult male leads sexual partner during consortship, or other chimpanzee is going to charge. Also seen in bonobos of Wamba (Kano 1998). (Type A) Category 1. See video.

look between thighs (LOT)
Quadrupedal individual lowers head and looks at another through spread thighs to solicit play. Mother uses to 'capture' infant as it continues to play when mother seeks to depart. Two-year-old female, Athena, looked at following observer between thighs at rear of group travel. Estrous young adult female presented to adolescent male, but he did not copulate. Then female looked at him through thighs in order to monitor his further response. (Type A) Category 7–8. See video.

look up (LOU)

Sitting or standing individual raises up face to monitor arboreal stimuli, e.g. fruit, colobus monkey, or others. (Type A) Category 1. See video.

Lopé

Study area of *Pan troglodytes troglodytes* in Lopé Faunal Reserve of Gabon (0°10′S, 11°35′E). Research organized by C.E.G. Tutin and M. Fernandez from 1984 to 1993 (Tutin et al. 1991).

loser support

See <support subordinate>.

lost call

See <whimper-scream>.

lost child

See <search for conspecific>.

louse

Ectoparasites of two genera, *Pediculus* and *Pthiris*, parasitizing chimpanzees. See <remove lice>. See Fig. 25.

Fig. 25 Louse: Louse of *Pediculus schaeffi* recovered from chimpanzee of M group (K. Zamma)

lower arm (LWA)
Groomee lowers arm, e.g., when grooming-hand-clasp ends, signaling intention to change role or posture. Groomer uses to request groomee to change role in bonobos (Kano unpublished). (Type A) Category 2–3.

lower head (LWH)
Two individuals face each other in social grooming, and one flexes neck to offer top of head for grooming. Also in bonobos (Kano unpublished). (Type A) Category 1–2. See video.

lower head and shoulder (LHS)
Mother lowers head and shoulders to drop off infant forward. See also <drop infant>. (Type A) Category 3–6. See video.

lower leg (LWL)
Resting or lying groomee lowers raised leg, apparently stimulating groomer to groom another body-part. Seen in bonobos (Kano unpublished). (Type A) Category 2. See video.

lower rump (LOR)
Mother lowers rump in order to drop infant. Sometimes, almost stands bipedally but with one hand on ground. See <drop infant>. (Type A) Category 2–7. See video.

Lui Kotale
Study area of *Pan paniscus* in Salonga National Park, DRC (2°45′S–20°22′E). Research organized by G. Hohmann and B. Fruth from 2000 to present (Hohmann and Fruth 2007).

Lukuru
Study area of *Pan paniscus* in Lukuru Forest, DRC (3°45′S, 21°05E). Research organized by J. Thompson from 1992 to present (Thompson 2002).

lunge
See <leap on>.

Luo Scientific Reserve
See <Wamba>.

M

Mahale

Research area of *Pan troglodytes schweinfurthii*, used as synonym of <Kasoje> area, and abbreviation of Mahale Mountains and Mahale Mountains National Park (6°07′S, 29°44′E). Research organized by T. Nishida from 1965 to present (Nishida 1990).

make bed (MBD)

Goodall (1989) Nest. Make platform for sleep or rest, by bending branches over each other. “Bed” preferable to “nest” because major function is for sleeping, not rearing. Includes <bend>, <transport>, <sit>, and <stand bipedal>. (Type B) Category 2. See video.

make cushion (MCU)

Bend or break one or few shrubs or other vegetation, then sit or lie on them above ground. Used for daytime rest or nap. Cushion is less complex construction than bed. Weanling may make cushion in order to be suckled. Customary at Mahale. Bonobos of Wamba (Kano 1998) and Lomako (Hohmann and Fruth 2003) do the same. (Type B) Category 2. See video.

make day bed (MDB)

Goodall (1989) Day nest. Bed made in daytime. Simpler in construction (Hiraiwa-Hasegawa 1989) and made more quickly (usually less than 1 min) than night bed. Bonobos make day beds (Kano 1998). See also <make cushion>. (Type B) Category 2. See video.

make ground bed or cushion (MGB)

Make bed or cushion on ground for daytime rest, by bending over few shrubs or grasses. Mahale chimpanzees make no ground beds for overnight use, but those of Nimba sometimes spend night on ground bed (Matsuzawa and Yamakoshi 1996) probably because of lack of predators, or as mate-guarding (Koops et al. 2007). Also, ground beds seem to be used for overnight sleeping at Kalinzu, with few large carnivores (Furuichi and Hashimoto 2000). (Type B) Category 3. See video.

make leaf cushion (MLC)

Hirata et al.'s (1998) Leaf cushion: Use leaves as cushions to sit on wet ground. Whiten et al.'s (1999) Seat vegetation. Different from simple cushion in that materials for leaf cushion is cut off from plant. Habitual at Taï, present at Bossou, Goualougo and Kibale (Sanz and Morgan 2007). Absent at Mahale. (Type B). See Fig. 26.

Fig. 26 Make leaf cushion: Leaves left by chimpanzee at Bossou (By courtesy of Satoshi Hirata)

make night bed (MNB)

Goodall (1989) Night nest. Plooij's (1984) Nestbuilding (NES). Evening bed-making is common characteristic of great apes. Durable sometimes for more than 6 months (Ihobe 2005) (Type A) Category 1–2. See video.

make sound (MSD)

Produce audible sound by exerting force on external object. Includes <clip leaf>, <drum>, <rap>, <slap>, <splash water>, <stamp>, etc. Cf. <vocalize>. (Type D) Category 1.

make tool (MTO)

Modify natural object, so that it serves more effectively (Beck 1980). At Mahale, this is limited to a few patterns. For example, make probe for fishing for carpenter ants: peel bark and cut it to appropriate size, use twig by removing leaves and small branch, remove leaf-blade and use mid-rib, etc. Also bend branches or shrubs for bed-making and courtship. Many other tool-making behavioral patterns seen at other sites. (Type D) Category 1. See video.

male invite

See <open thighs>.

marrow-pick

See <pick out bone marrow>.

mass excitement (MAE)
Nishida (1970); Mori's (1982) Booming situation; Reynolds and Reynolds' (1965) Carnival; Goodall's (1989) Social excitement. Simultaneous, noisy display, when many chimpanzees gather at one place, such as new food source. (Type D) Category 3.

massage shoulder (MAS)
When trouble started and screaming began, adult female, Xtina, massaged shoulder of female, Ruby, with left hand, while both pant-hooting. Kind of reassurance? (Type A) Category 8–9. See video.

masturbate (MAB)
Includes <fumble penis> for males and <fumble clitoris> for females. (Type B) Category 1–8.

mate
See <copulate>.

mate guarding
See <herd>.

maternal behavior
See <care maternally>.

maternal leave
See <travel alone after childbirth>.

mature
See <adult>.

meat-eating
See <eat meat>.

meat-sharing
See <share food>.

medicate self (MDS)
Includes <lick wound>, <lie>, <remove lice>, <remove sand flea>, <push peri-anogenital region with finger>, and <probe nasal passage>. <Swallow leaf> and <chew> *Vernonia* piths may be internal self-medication (Huffman and Kalunde 1993; Huffman 1997). (Type D) Category 1.

menstrual cycle
See <swelling of sexual skin>.

misunderstand (MIS)
Misconstrue other's intention, e.g. flinch from accidentally dropped stick. Adolescent female, Ai, tried to escape when adult female, Xtina, approached to groom her. Ai misunderstood that Xtina would attack her. (Type D) Category 1–3. See video.

mock bite
See <mouth>.

monitor (MNT)
Investigate changes in environment or social interactions. Includes <watch>, <look up>, <stand bipedal>, <walk bipedal>, <sniff>, etc. (Type D) Category 1. See video.

monitor monkeys (MNC)
Chimpanzees in "hunting mood" sit on ground, scan canopy, and walk little by little, perhaps to look for immature monkeys and to check general movements. Includes <walk>, <run>, <climb>, <watch>, <look up> etc. (Type D) Category 3. See video. See also video "hunt."

monitor mother (MNM)
Infant being weaned by rejecting mother runs screaming from her to few meters away, then repeatedly glances at mother to see if she will allow suckling, or cradling. Includes <whimper>, <scream>, <whimper-scream>, <leave to protest>, <run>, <look back>, <glance>, and <throw temper tantrum>. Also in bonobos (Kano unpublished). (Type D) Category 1–2. See video.

mop ant (MOP)
Collect *Camponotus* ants with hair on back of hand, usually when many wood-boring ants emerge from nest and move about on surface of tree trunk. Small infant may use palmar side of hand. (Type A) Category 7. See video.

M

mop ant with leaves (MOL)
Wipe away *Camponotus* ants with crumpled leaves. Observed only once for K group female, Chausiku. (Type A) Category 9.

moralistic aggression
See <aggress morally>.

mother-offspring relations
Mother and offspring spend most time together throughout infancy and juvenility. Since son remains in natal group throughout life, mother-son relationship lasts until either of them dies. By contrast, since daughter usually emigrates from natal group and never returns, mother-daughter relationship terminates abruptly around 11 years of daughter's age.

Moukalaba
Study area of *Pan troglodytes troglodytes* in Moukalaba-Doudou National Park, Gabon (2°33′S, 10°57′E). Research organized by J. Yamagiwa from 2003 to present (Takenoshita et al. 2008).

mount (MOU)
Embrace another with arms from behind. Goodall (1989): "Part of the ventral surface of the mounter is in contact with part of the dorsal surface of

the other, and the mounter leans forward over the other, usually grasping him/her". She also mentions that "the mounted individual (male or female) may reach back to touch the genital area of the mounter with a hand, (or, rarely, the foot), or the mounter may raise his/her own foot to the scrotum of the other." Goodall (1989) lists as synonyms: "Dorso-ventral embrace", "Mounting embrace" or "Mount-embrace". Perhaps, Kortlandt's (1964) Sham copulation. Plooij's (1984) MOU, de Waal's (1982) Mount. Males and females mount. Mounting occurs, especially among adult males, during high social excitement such as reunion of rival males, in which each rival male mounts adult male or female, and both participants often scream or pant-scream. Usually bipedal with pelvic thrusts. Appears to function as self-reassurance, reassurance, recruitment of coalition partners, etc. In bonobos, mounting occurs in similar contexts such as appeasement and easing of tension, but posture differs, so that mounted individual usually stands quadrupedal (Kano 1998). (Type A) Category 2. See video.

mount in copulation (MOC)
Goodall (1989) Copulation mount: "Male places both hands firmly on the female's back or sides, during which he may open mouth kiss her. … accompanied by intromission and thrusting movements of the pelvis." (Type A) Category 1. See video.

mount, misdirected (MOM)
van Hooff (1973) Disoriented mounting: "Thrusting towards the head, face, shoulder or any other part except the rear." (Type A) Category 3. See video.

mouth (MOT)
van Hooff (1973) Gnaw: "Keep month wide open without the retraction of lips and press the teeth to the back, shoulder or other body part without biting." Reassurance often accompanied by panting. Similar pattern used to tickle with mouth pressed to shoulder, chest, belly, etc. Nishida's (1983a) Mouthing or Mouth contact. Plooij's (1984) Mouthing (SAB). Goodall's (1989) Biting includes <mouth>. Goodall's (1989) Play biting occurs in social play. See also <play bite>. Cf. <bite>. Kano's (1998) Mock bite for bonobos. (Type A) Category 1. See video.

mouth and shake (MTS)
Stand quadruped, mouth and shake another continuously, in social play. (Type B) Category 7.

mouth for begging (MOB)
Beg for food by putting lips to lips or hand of feeding other. Goodall's (1989) Beg mouth-to-mouth. Plooij's (1984) Beg with mouth (BWM). Kano's (1998) Food beg: mouth to hand and mouth to mouth. Cf. <extend hand to beg>. (Type A) Category 1. See video.

move (MOV)
Dislodge object such as rock, heavy branch, etc. by <push>, <pull>, <lift>, <stamp>, etc. Pika's (2007) Move in bonobos. (Type D) Category 1. See video.

mummify (MUM)

Mother sometimes keeps dead infant for months, so it may be mummified if death occurs in dry season (Nishida 1973b). Also reported from Bossou (Matsuzawa 1997). (Type D) Category 3. See Fig. 27.

Fig. 27 Mummify: Adult female, Wakasunga, continued to carry corpse of her infant until it was mostly bare bones (T. Nishida)

mummy

See <mummify>.

mutual grooming

See <groom mutually>.

muzzle-rubbing

See <rub muzzle>.

N

nasal probe

See <probe nasal passage>.

Ndakan

Study area of *Pan troglodytes troglodytes* in Dzanga-Sangha region of Central African Republic (2°20′N, 16°10′E). Research done by J.M. Fay in 1988 (Fay and Carroll 1994).

Ndoki

Study area of *Pan troglodytes troglodytes* in Nouabale-Ndoki National Park, Republic of Congo (2°20′N, 16°19′11″E). Research organized by S. Kuroda from 1989 to 1992 (Kuroda et al. 1996).

neck pocket

See <transport in neck pocket> and <hold object in neck pocket>.

neglect

See <snub>.

nest

See <bed> and <make bed>.

nest-build

See <make bed>.

nest-grunt

See <grunt in bed>.

nest play

See <play in bed>.

Ngogo

Study area of *Pan troglodytes schweinfurthii* in Kibale National Park, Uganda (0°30′N, 30°25′E). Research organized by J.C. Mitani and D.P. Watts, continuing from 1994 to present (Mitani and Watts 1999). Pioneer study done by M.P. Ghiglieri in 1976–1978 and in 1981 (Ghiglieri 1984).

Ngotto Forest

Study area of *Pan troglodytes troglodytes* in Central African Republic (4°01′N, 17°04′E). Research organized by R.S. Fouts from 2000 to present (Hicks et al. 2005).

night nest

See <make night bed>.

Nimba

Study area of *Pan troglodytes verus* in Nimba Mountains Biosphere Reserve, in Guinea and Cote d'Ivoire (7°38′59″N, 8°25′24″W). Research done intermittently from 1991 to present by Y. Sugiyama and colleagues, especially K. Koops and T. Humle (Koops and Matsuzawa 2006).

nipple contact

See <contact with nipple>.

nipple press

See <fumble nipple>.

nod and mouth (NOM)

When seated, holds playmate (often infant) in lap, or as it stands quadrupedal, repeatedly brings mouth to body of playmate, in nodding motion. Usually elicits play pants from recipient. Adult male hugs another from back and repeatedly puts open mouth to latter's back for self-reassurance. (Type A) Category 7. See video.

N

nod to water surface (NOW)

Nod up and down repeatedly while standing quadrupedal and watching reflection of self on water's surface. Few individuals at Mahale such as Cadmus, Michio, Jiddah and Totzy do so (Nishida et al. 2009). Cf. <look at water>. (Type A) Category 8. See video.

nod to water surface and mouth water (NWM)

Adult female, Totzy, nods to water's surface. She repeatedly takes water into mouth when face down and spits it out when face up. Rare water-contact play. (Type A) Category 9. See video.

nod with body part in mouth (NOB)

Juvenile male, Xmas, sits and nods repeatedly with hand of infant playmate in mouth. Rare social play pattern. (Type A) Category 8. See video.

nod with object in mouth (NOO)

Juvenile female, Jiddah, puts pebble from stream into mouth, retains it and nods repeatedly. Idiosyncratic solo play. (Type A) Category 8. See video.

nod with play face (NOP)

Nod up and down repeatedly, with play face, while quadrupedally standing or approaching playmate. Solicitation of play. (Type A) Category 7–8. See video.

nut cracking

See <hammer nut>.

nut-hammer

See <hammer nut>.

nuzzle (NZL)

Goodall (1989) Nuzzle: "The young infant, until about 3 months old, moves its head from side to side and up and down against the mother's body when searching for the nipple…" Termed rooting behavior in other mammalian species. Kano's (1998) Nuzzle in bonobos. (Type A) Category 1.

O

object play
See <play with object>.

Odzala
Study area of *Pan troglodytes troglodytes* in Odzala National Park, Republic of Congo (0°23′–1°46′N, 14°16′–16°40′E). Research done by C. Devos and colleagues in 2001 (Devos et al. 2002).

offer arm (OFA)
Dominant individual extends arm for subordinate to mouth, to be reassured, or to be groomed. Absent in bonobos of Wamba (Kano unpublished). (Type C) Category 3–5. See video.

offer back (OFB)
Present back to solicit grooming of it. (Type A) Category 1–2. See video.

O

Okorobiko
Study area of *Pan troglodytes troglodytes* at Mt. Okorobiko of Rio Muni, Equatorial Guinea (1°–2°30′N, 9°30′–11°30′E). Research done by C. Jones and J. Sabater Pi (1969, 1971) from 1966 to 1968.

old age
See <senescence>.

open eyes (OPE)
Plooij (1984) Eyes open (EYO). (Type A) Category 1. See video.

open mouth
See <play face> and <yawn>. Some individuals open mouth when showing relaxed face.

open mouth kiss
See <kiss with open mouth>.

open thighs (OPT)
Goodall (1989) Male invite: "…male sits with thighs splayed and penile erection, looking toward a female in estrus. Sometimes his hair is erect…"

Male courtship display. Similar pattern in bonobos (Kano 1998). (Type A) Category 2. See video.

orphan

Dependent offspring who has survived mother's death. Usually accompanies adult males, particular adult female, or sibling. See <adopt>.

ostracize (OST)

Most members of unit group punish or expel individual from group by repeated, concerted attack. Reported first by Nishida et al. (1995). See <kill adult male> and <attack concertedly>. (Type D) Category 1–5. See Fig. 28.

Fig. 28 Ostracize: Ex-alpha male, Ntologi, spent about 8 months alone after being ostracized (T. Nishida)

P

palmigrade walk

See <walk quadrupedal on palms>.

pant (PAN)

Goodall (1989): "…various calls which are linked by audible inhalations and which may be described as vocalized pant. Non-vocal panting sometimes occurs when chimpanzees are grooming each other, during greeting, etc. Sometimes, when the mouth is closed, only breathing sounds are heard. The best indication of panting is the quick rhythmic movements of the body that accompany it." Bonobos pant in play, but not in other contexts? (Type B) Category 2–3. See video.

pant-bark (PBA)

Goodall (1989): "Series of bark-like sounds joined by voiced inhalations." See also <bark>. Status unknown in bonobos. (Type A) Category 2–3. See video.

P

pant-grunt (PGR)

Goodall (1989): "Series of soft or loud grunts functioning as a token of respect given during greeting by submissive chimpanzees and during submissive interactions… A highly fearful individual may utter frenzied pant barks that may be labeled pant-screams." van Hooff's (1973) Rapid OhOh. At Mahale, pant-grunting adult females may <present> to adult males, who may mount her and pelvic thrust. In response to pant-grunting from young adult and adolescent males, alpha male may jump on or attack them without being appeased. See also <bob> and <bow>. Pant-grunt absent in bonobo (Nishida and Hiraiwa-Hasegawa 1987). (Type A) Category 3. See video.

pant-grunt with bent elbow (PGE)

Pant-grunt while crouched quadrupedally with elbow flexed. Often, part of pant-grunt complex performed by older juvenile and adolescent males at Mahale, although adult females also do so. Cf. <bow>. (Type A) Category 3–5. See video.

pant-hoot (PHT)

Goodall (1989): Series of long calls "which are often contact calls between distant groups or calls given by chimpanzees at night, from their nests, when they are within earshot of another group… give pant-hoots when they arrive at a food source, cross a ridge, or face into a new valley. A subordinate greeting a dominant often gives pant hoots and pant grunts…" Higher-ranking males pant-hoot more often than lower-ranking ones (Clark 1993; Mitani and Nishida 1993). Chimpanzees advertise their presence and numerical strength to rival groups (Wilson and Wrangham 2003). At Mahale adult males also pant hoot when agonistic confrontations are continuing, in which pant hoots function as vocal threat (Nishida 1983b). Pant-hoots have four stages: introduction, build-up, climax and let-down, but females lack climax and let-down stages (Mitani and Nishida, 1993). Elements of pant-hoot calls vary between Mahale and Gombe (Mitani et al. 1992). Exact bonobo version of pant-hoot absent (Nishida personal observation), but Kano's (1998) Waa call, or Okayasu's (1991) Hoot may be homologous to pant-hoot. Bonobo Waa call emitted upon arrival at feeding site, during feeding and resting, and at sunset after making night beds (Kano 1998). (Type A) Category 2. See video. See also videos "drum", "pat" and "pull rock to roll".

pant in copulation (PAC)

Goodall (1989) Copulation pant: "Rather hoarse panting given by some males during copulation. In its loudest form may be confused with laughing." At Mahale rarely heard and seems idiosyncratic.
(Type A) Category 8.

pant-scream

See <pant-grunt>.

P

panting-laugh

See <play-pant>.

paraplay

See <solicit play> and <rebuff play>.

parry (PAR)

van Hooff's (1973) Parry: "One or both arms are raised. The forearm is kept in a roughly horizontal position over or in front of the head, thus shielding it off from possible beats from a fellow." Goodall's (1989) Startle flinch: "When a chimpanzee is startled by a sudden movement nearby (such as a low-flying bird, large insect, unexpected gesture of human or even another chimp, etc.) he/she will immediately duck his head and fling one or both arms across his/her face, or if very startled, throw both hands in the air…" Also, Goodall's (1968) Startle reaction;

Plooij's (1984) Startle (STT). Bonobos of Wamba duck head in similar context (Kano unpublished). (Type A) Category 1. See video. See also video "splash water."

party

Sugiyama (1968) Party. Nishida (1968) Subgroup. Temporary subgroup of unit group or community, in fission-fusion. Size of parties changes seasonally, dependent on distribution and quantity of major fruit (Itoh and Nishida 2007). See <unit group>.

pass (PAS)

Walk past another, without body contact. See <walk quadrupedal>. Goodall's (1989) Pass. Kano's (1998) Pass without body contact for bonobos. (Type C) Category 1. See video.

pass under (PAU)

Go under low horizontal obstacle such as fallen log. Infant may walk quadrupedal under adult walking or standing, perhaps for amusement. Juvenile female, Imani, once passed through between legs of human (TM). (Type C) Category 8. See video.

pass with body contact (PAB)

Pass by another and make unnecessary body contact (with shoulder, not hand or foot), apparently intentionally. Two young adult males and three juvenile females at Mahale did so to human observer in early 2000s. Significance unknown. Kano's (1998) Pass with body contact for bonobos. (Type B) Category 2. See video.

pat (PAT)

Plooij (1984) Pat. Goodall (1989): "Repeated brief contacts, with the palmar surface of the hand, on the body of another, usually on the head, back or hand. This is a reassurance gesture directed towards a submissive individual by a more dominant individual. It has a calming effect." Also directed to infant's head or back. Shown in play by juvenile bonobos of Wamba (Kano 1998). (Type A) Category 1. See video.

patrol (PTR)

Nishida (1979) Scout. Goodall (1989): "Party of males (occasionally accompanied by a female, usually in estrus) moves deliberately and silently in the peripheral part of community range." Elements include <sniff>, <stalk>, <stare fixedly>, <attack>, and <flee>. Also from Budongo (Reynolds 2005), Kanyawara (Wrangham and Peterson 1996), Ngogo (Watts and Mitani 2001; Mitani and Watts 2005) and Taï (Boesch and Boesch-Achermann 2000). Bonobo males silently approach adjacent group, but only after hearing latter vocalize. Kano's (1998) Scout. (Type D) Category 1–3. See Fig. 29.

Fig. 29 Patrol: Adult males of M group patrolled and invaded K group's range at Myako Valley. Two adult males lead travel (T. Nishida)

peel with hand (PEH)

Pull tip of outer layer of stalks of grasses, herbs and woody vines by hand while holding stalk with other. Supplementary movement of <peel with teeth>. Also in bonobos (Kano 1998). (Type A) Category 1. See video.

peel with teeth (PET)

Pull with teeth tip of outer bark while holding stalk in hands, thus removing hard outer layer of stalks of grasses (sugar cane and elephant grass), herbs (ginger and *Marantochloa*) and woody vines (*Landolphia* and *Saba*) to eat pith. Peeled bark used as probe for ant-fishing. Also in bonobos (Kano 1998). (Type A) Category 1. See video.

peep (PEP)

Look through hole or slit. Cf. <peer>. (Type A) Category 1. See video.

peer (PER)

Look intently into another individual's face from few cm. distance, usually in quadrupedal stance. Goodall (1989): when another eats, leaf-grooms, investigates or licks wound. In addition, at Mahale, when trying to remove sand fleas. Plooij's (1984) Put face close (PFC). Peering usually does not elicit food sharing. Cf. <glance>, <look>, <stare fixedly>, <watch>, and <sniff mouth>. Bonobos show same pattern (Idani 1995; Kano 1998; Pika 2007). (Type A) Category 1. See video.

peer together (PEG)
Up to 11 individuals gather, encircle, and peer at targeted individual who attracts attention. Peered-at has new wound, or is removing ectoparasites. Recorded also at Bossou (Sugiyama 2008). (Type B) Category 1–3. See video.

pelvic thrust
See <thrust>.

penile adduction
See <adduct penis>.

penile erection
See <erect penis>.

perforate (PFR)
Chimpanzees of Goualougo "reopen termite exit/entry holes on the surface of epigeal nests after unsuccessfully attempting to open the holes by hand. Perforating twig tools are held with a precision grip as the tip is pressed into the surface of the mound. The end of the tool is used to push through the soil replaced by termites to seal the nest" (Sanz et al. 2004). Absent at Mahale. (Type B) See video (Goualougo).

pestle-pound
See <pound pestle>.

Petit Loango
Study area of *Pan troglodytes troglodytes* in Petit Loango Faunal Reserve, Gabon (2°20′S, 8°0′E). Research done by J. Yamagiwa and colleagues from 1994 to 1998 (Takenoshita et al. 1998).

pick ear (PCE)
Remove ear wax with finger. See <sickness>. (Type A) Category 1–3.

pick nose (PCN)
Remove nasal mucus with finger and usually eat it. See <eat nasal mucus> and <sickness>. (Type A) Category 1. See video. See also video "eat nasal mucus."

pick out bone marrow (PCB)
Marrow-pick (Whiten et al. 1999): Put stout probe into cracked bone and extract bone marrow with it. Customary only at Taï. Present at Goualougo (Sanz and Morgan 2007). Absent at Mahale. (Type B)

pick out nutmeat (PCA)
Chimpanzees of Bossou pick out residual contents of nutshell with probe after cracking open oil palm fruit and eat contents (Sugiyama 1997). Present at Goualougo (Sanz and Morgan 2007). Absent at Mahale. (Type B) See Fig. 30.

Fig. 30 Pick out nutmeat: Drawn from photo of Bossou chimpanzee (By courtesy of Michio Nakamura)

pick out pulp (PCP)

Pick out with forefinger pulp from large fruit such as *Saba comorensis* and eat it. (Type A) Category 1. See video.

pick up (PIU)

Take object such as food by hand from substrate such as ground. Cf. <lift>. (Type A) Category 1. See video. See also video "search for object."

pick up and release (PCR)

While sitting, grasp and lift objects, such as dry leaves, sand, pebbles, etc. and release in air. Solo play in infants. (Type B) Category 1–8. See video.

pick up discarded food (PDF)

Nishida and Turner (1996) Food retrieval. Pick up food left by others. Infant often picks up food left by mother and ingests it, thus gradually learning what to eat. (Type C) Category 1. See video.

pick up lice

See <remove lice>.

pile dry leaves (PLE)

Spread arms to right and left sides and bring together dry leaves in front of body by adduction of arms. Produces distinctive rustling sound. Solo play by youngsters. (Type A) Category 1–7. See video.

pin down

See <press down>.

pinch aggressively (PIN)

Goodall (1989) Pinch. Plooij's (1984) Pinch (PIN). Squeeze skin between thumb and index finger. Idiosyncracy of male chimp at Gombe. Absent at Mahale. Absent in bonobos (Kano 1998). (Type A)

pinch clitoris (PIC)

Pinch clitoris of juvenile or infant female as tickling. (Type B) Category 8. See video.

pirouette (PIR)

Goodall (1989): "… (usually a youngster) progresses in a series of tight circles, moving on all fours or on their feet. Type of locomotor play." Seen in older infants, juveniles and young adolescents at Mahale. Juvenile female, Ivana, once pirouetted in order to solicit play from small infant, but usually done as solo play (Nishida and Inaba 2009). Probably absent in bonobos of Wamba (Kano unpublished), perhaps different from Pika's (2007) Ice skating. (Type A) Category 3–6. See video.

play (PLY)

Goodall (1989) divided play into lone (or solo) play and social play. She split play into four categories, (a) locomotor play, or van Hooff's (1973) Gymnastic, (b) nest play, (c) object play, (d) self play. She added that, "There is a facial expression connected with play, the play face, a type of locomotion that is seen only in the play context, the play walk, and a vocalization, laughing." Goodall (1989) listed as elements of each play type: (a) locomotor play: <dangle>, "gambol", <swing>, <pirouette>, and <somersault>. (b) nest play: <make bed>, <leap>, and <climb>. (c) object play: "pick up", <throw>, "rub", <rotate fruit>, <drag>, <drape>, <flail>, and <move>.(d) self play: <grab>, <dangle>, <slap>, and <tickle>. Social play included <play bite>, <push head into ventral>, <chase>, <circle>, <drag>, <wrestle with fingers>, <hang-wrestle>, <kick heel>, <kick>, <kick back>, <pinch>, <play walk>, <pull>, <push>, <step on>, <rub>, <slap>, <spar>, <tickle>, <pull object from opposing sides>, <wrestle>, etc. in addition to most patterns used in lone play. (Type B) Category 1. See video.

P

play bite (PLB)

Goodall (1989) Play bite: "Contact made on the partner's body with teeth. Sometimes one of the players gets a firm grip on a limb and holds on for minutes at a time…" Infant, Teddy, bit mother's supra-orbital region, etc. softly as play. See <mouth>, as play bite is <mouth> in play. (Type A) Category 1–2. See video.

play face (PFA)

van Hooff (1972) Relaxed open-mouth display. Goodall (1989): "Only slight retraction of the lip corners. Jaws may be closed or open… Occurs during play." Also in bonobos (Kano 1998). (Type B) Category 1. See video.

play face, full (PFF)

Goodall (1989) Full play face: "Upper and lower teeth show as the play gets rougher and tickling more vigorous…" Plooij's (1984) Play face full (PFF). Also in bonobos (Kano unpublished). (Type A) Category 1. See video.

play face, half (PFH)

Plooij (1984), Goodall (1989): Low play face: "Only the lower teeth show." Status unknown in bonobos. (Type A) Category 1–3. See video.

play fight

See <wrestle>.

play, imaginary (PLI)

Solo play, apparently with imaginary companion or tool. Multiparous adult female lay supine with all limbs raised and play-panted as if playing "aeroplane" with infant who had been dead for few months (Nishida et al. 1999). Adolescent male ran round and round tree while play-panting as if chasing, or fleeing from playmate (Hayaki 1985a). Adolescent male of Kanyawara carried log for over 4 h, treating it as infant (Wrangham and Peterson 1996). Rarely observed. See also Gomez and Martin-Andrade (2005). (Type D) Category 1–8.

play in bed (PLN)

Goodall (1989) Arboreal nest play. Also in bonobos (Kano 1998). (Type B) Category 2. See video.

play invitation

See <solicit play>.

play-pant (PLP)

Plooij's (1984) Laugh (LAC). Goodall (1989) Laugh: "Somewhat resembles human laughter, heard when chimpanzees are playing socially. As the play becomes rougher, or involves more tickling, there is more laughter…" Chimpanzees rarely emit play-pant during solo play. In social play, tickled or chased chimpanzee emits most play-pants (Matsusaka 2004). Bonobos also play-pant (de Waal 1988; Okayasu 1991; Kano 1998). (Type A) Category 1. See video. See also video "hang with legs pitterpat."

play, parallel (PAP)

Two or more infants play simultaneously in vicinity, but without interacting with one another. (Type D) Category 1. See video.

play, rough and tumble (PRT)

Most physical type of play, usually between youngsters. Paquette's (1994) Rough play. Wrestling and often locomotor play. Elements include <chase>, <run>, <climb>, <leap>, <grab>, <tag>, <wrestle>, <push>, <pull>, <mouth>, <play bite>, <drag>, <stamp>, <slap>, <thrust>, <play-pant>, <scream>, <play face>, <grin>, etc. (Type B) Category 1. See video.

play self
See <play solo>.

play socially (PLA)
Goodall (1989) Social play. Play with conspecific. Usually in twos, but sometimes in 3–5. See also <play>. (Type D) Category 1. See video.

play socially with object (PSO)
Goodall (1989): Social object play. Social play that includes chasing individual who has object such as fruit, branch, stone, animal skin, etc. and simultaneously pulling object from opposing sides. At Mahale, youngster may also press leafy branch to playmate while wrestling. Ingmanson's (1996) Tool use in play for bonobos. (Type B) Category 1. See video. See also video "discard."

play solo (PLS)
Goodall (1989) Self play. Mendoza-Granados and Sommer's (1995) Solitary play. Play alone. Cf. <play socially>. (Type B) Category 1. See video. See also video "enter hole".

play start
See <solicit play with object in mouth>.

play walk (PWL)
Goodall (1989): "The chimp walks with a rounded-back, its head slightly bent down and pulled back between the shoulders while it takes small stilted steps." See also <tilt head>. Plooij's (1984) Play walk (PLW). Seen in bonobos (Kano unpublished). (Type A) Category 2. See videos.

play with another animal (PWA)
Play with animal of another species. Includes wrestle and chase-flee play. At Gombe chimpanzee youngsters play with olive baboons (Teleki 1973). At Mahale chimpanzees from older infancy to adolescence seem to solicit human observers to play with them by <hit and run>, <somersault>, <lift and drop>, etc. Bonobos of Wamba play with red-colobus monkeys (Ihobe 1990). Cf. <care alloparentally for another species> and <trifle with>. (Type D) Category 8.

play with object (PLO)
Goodall (1989) Object play. Solo play, handling object such as fruit or stone. See also <play>, <play socially with object> and <toy>. (Type B) Category 1. See video.

play with sand (PYS)
Sand apparently stimulates youngsters of M group to lone or social play. They roll over sand or somersault on it. Repeatedly grab, raise and release handfuls of sand in sitting or quadrupedal posture, appearing to enjoy movement and sound of sand. Not reported from elsewhere. (Type B) Category 1–7. See video.

play with urine (PLU)
Infant touches with hand urine passed by another chimp sitting overhead. (Type B) Category 1–7. See video.

play with water (PLW)
Solo play including <splash water>, <look at water>, <stir water>, <move> rock, <nod quadrupedal> etc. (Type B) Category 1–7. See video.

poke (POK)
Nishida (1983a): Tickle, by prodding repeatedly with fingertips, with one or more extended fingers to infant's ventrum. Adult also poke dominant individual's hand rhythmically for reassurance. Goodall (1989) mentioned that "Sometimes an individual may poke another chimp who has ignored a solicitation for grooming." Plooij's (1984) Poke (POK). Seen in bonobos. (Type A) Category 1. See video.

poke in social scratch
See <scratch socially, poke type>.

poke with play face (POP)
Poke once another individual with one finger, while showing play face. Type of soliciting play. (Type B) Category 1–7. See video.

police (PLC)
Type of impartial intervention by adult males. Dominant males rush to place of commotion and run past quarreling parties, in order to stop quarrel. Boehm's (1999) "control role" by alpha male at Gombe. (Type D) Category 3. See videos.

possessive behavior
See <herd>.

post-conflict behavior
See <aggress, redirected>, <appease>, <console> and <reconcile>.

pound (POD)
Sanz and Morgan (2007): "percussion of tool onto another object or substrate". Pounding beehive customary at Goualougo (Sanz and Morgan 2007), and likely to occur at Bwindi (Stanford et al. 2000). Absent at Mahale. (Type A)

pound food on object (PDO)
Whiten et al. (1999): Food-pound onto substrate, e.g. stone, root, etc. Customary at Gombe and habitual at Taï. Absent at Mahale. (Type B)

pound food on wood (PDW)
Whiten et al. (1999) Food-pound onto wood: smash food. Customary at Bossou, Taï and Gombe. Absent at Mahale. (Type B)

pound insect (PDI)
Whiten et al (1999) Insect-pound: Use object to mash insect. Present at Bossou. Absent at Mahale. (Type B)

pound pestle (PDP)

Yamakoshi and Sugiyama (1995) Pestle pound: Bossou chimpanzee climbs to crown of oil palm (*Elaeis guineensis*) and mashes apical growth tip of crown with mature petiole (frond) detached from oil palm. Unique to Bossou. (Type B) See video (Bossou).

pound resin (PDR)

Whiten et al. (1999) Resin-pound: extract resin by pounding. Present at Bossou. Absent at Mahale. (Type B)

pout (POU)

Plooij (1984) Pout face (PTF). Goodall (1989) Pout face: "Lip corners pushed forward, mouth may be slightly open. Lips together, except at the front where they may funnel out somewhat. This is seen with the hoo call and hoo whimper". Seen in bonobo (Kano 1998). (Type A) Category 1. See video.

pout face

See <pout>.

pout kiss

See <kiss with pout face>.

predation on chimpanzees

See <respond to predator>.

pregnancy

Females in last stage of pregnancy often <lie prone>, <rest>, and <travel> alone. Pregnant female often recognizable from large belly. Cf. <travel alone after childbirth>. See <gestation period>.

presbyopia

Farsightedness from aging presumed from greater distance old chimpanzee maintains from body part that he/she is grooming. See video.

present for grooming

See <solicit grooming>.

present with limbs extended (PRE)

Normal quadrupedal stance with hindquarters directed to dominant chimpanzee. Goodall's (1989) Present, also Rump-turn. Plooij's (1984) PRE. van Hooff's (1973) Mount-present. Kano's (1998) Dorso-ventral presenting for bonobos. Bonobo females also show ventro-ventral courtship gesture. (Type A) Category 2. See video.

present with limbs flexed (PRF)

van Hooff (1973) Crouch-present. Quadrupedal posture with limbs flexed, and hindquarters turned towards another. Differs from <crouch>, in which hindquarters not directed to partner. (Type A) Category 3. See video.

press (PRS)
Force contact with shoulder, side or hip against another. Function unclear, but may initiate social play or aggression. (Type A) Category 3–7.

press down (PRD)
Restrain another under body to immobilize partner against substrate. Common in social play, such as <wrestle> and sometimes aggression. Cf. <immobilize>. (Type A) Category 1–6. See video.

press neck with lower arm (PNA)
Press neck of wrestling partner with lower arm. (Type A) Category 8. See video.

press nipple
See <fumble nipple>.

press object on (PRO)
Press branch, herb, etc. against playmate, thus indirectly applying force. Adolescent and juvenile males, e.g. Cadmus and Michio, used in play. (Type B) Category 8. See video.

press teeth against back (PTB)
Adult male mounts another male, and mounter presses upper row of teeth against mounted's back, while mounter embraces other, thrusts pelvis, and has grin-full-open. Intensified form of <mouth>. <Mount> accompanied by <nod and mouth>. Functions as reassurance and coalition formation, and sometimes during copulation. Rarely, juvenile male presses teeth against objects, such as mass of dry leaves. Absent in bonobos of Wamba. (Type A) Category 3–6. See video.

probe (PRB)
Use vine, stem, bark, grass, twig, or stick to touch object. Goodall's (1989) Investigation probe. Infant chimpanzees appear innately inclined to push investigatory probe into tree hole, long before starting to fish for arboreal ants. Juvenile sometimes inserts probe inside of large fruit such as *Voacanga* to try to obtain juice. Customary at Gombe, Mahale, Bossou, and Taï. Present at Goualougo and Kibale. Probing for bees is customary at Taï and present at Mahale. Absent in bonobos of Wamba (Kano 1998). (Type A) Category 3. See video. See also video "probe with finger."

probe nasal passage (PNP)
Adult male, Kalunde, inserted probe into nostril to stimulate sneeze, which discharged much nasal mucus (Nishida and Nakamura 1993). He showed this behavior for decade (Nishida et al. 2009). Only one other member of M group, Maggie, female adolescent, did so (Marchant and McGrew 1999). (Type B) Category 8. See video.

probe with finger (PWF)
Infant inserts finger into small tree hole, etc., to investigate contents of cavity. Cf. <grope>. (Type A) Category 1–3. See video.

process food (PRC)

Change raw material of food for ingestion by removing inedible parts, rendering into manageable bits with hand and teeth, etc. Include <break branch with use of foot>, <break open by pushing chin>, <pull through with hand>, <pull through with mouth>, <twist>, etc. (Type D) Category 1.

prop anvil (PRP)

Whiten et al. (1999) Anvil-prop: Insert smaller stone under anvil to level its working surface. Seen multiple times at Bossou (Matsuzawa 1994; Biro et al. 2006). Unique to Bossou. (Type A)

protect (PTC)

Goodall (1989) Protect. Caretaker protects infant from rain, storm, danger (e.g., gap in canopy), or social harassment (e.g., male charging display). Also in bonobos, except for protection from rain (Kano 1998). (Type D) Category 1. See video.

protest (PTT)

When attacked, threatened, or stolen from, resolutely display dissatisfaction with <bark>, <stare fixedly>, <scream>, and <chase> to dominant attacker, especially if supporter present, then attacker may flee. Supporter of victim (often mother of infant or juvenile victim) may join in or remain calm. Cf. <retaliate>. (Type D) Category 1–3. See video.

protrude tongue (PTG)

Small infant sticks out tongue. Function unclear. (Type A) Category 3–8. See video.

pucker cheek (PCC)

Suck in cheeks. Function unknown. (Type A) Category 3–8. See video.

P

pull (PUL)

Grasp and tug at object or other by flexing arms. Often in feeding, play, grooming, or agonism. Alpha male of M group pulled estrous female unwilling to follow him. Estrous female, Ako, pulled juvenile grooming adult male so vacating place for her to copulate with him. Plooij's (1984) Pull (PUL). Kano's (1998) and Pika's (2007) Pull for bonobos. (Type A) Category 1. See video.

pull down (PLD)

Stretch arm to reach upper branch of shrub, often while standing bipedal, and pull branch downward. Then, sit and eat fruits or leaves while gripping branch with hand. Mother or playmate pulls down infant from overhead. Also in bonobos (Kano unpublished). (Type A) Category 1. See video.

pull each other (PEO)

Two individuals simultaneously pull from opposite sides. Social play rarely observed. Two adolescent chimpanzees once pulled each other at stream, one standing in water and other standing on stepping stone. (Type B) Category 8–9. See video.

pull face to face (PLF)
Pull hand or other body part of another, to face each other. Occurs when chimpanzee wants to groom other's venter. Also in bonobos (Kano unpublished). (Type C) Category 1–2.

pull head with hands (PLH)
Pull head of another with hands, often in grooming. Also in bonobos (Kano unpublished). (Type A) Category 1–2. See video.

pull leaf-pile (PLL)
Walk backwards quadrupedally while raking dry leaves with hands. Locomotor-object play (Nishida and Wallauer 2003). Cf. <push leaf-pile>. (Type A) Category 7. See video.

pull object from opposing sides (POS)
Tug of war. Two individuals (or, rarely, two groups of individuals) simultaneously pull object (infant, branch, carcass etc.) competitively from opposite sides. Component of social play. Also in bonobos (Kano unpublished). Cf. <pull each other>. (Type B) Category 1. See videos. See also video "play socially with object".

pull out (PUO)
Pull forcefully with hands, often countered by pushing feet against ground. Chimpanzee tries to uproot stalk of banana, elephant grass, or African ginger from earth. Also, standing quadrupedal and pulling up heavy rotten wood with arms. Also in bonobos. (Type A) Category 1. See video.

pull rock to roll (PUR)
Adult or adolescent male pulls rock along substrate with hand while walking or running, as part of charging display. Immature pulls or rakes rock as solo play while squatting. Same movement to pull up heavy wood with arms. (Type A) Category 2–7. See video.

pull through in courtship (PTS)
Nishida (1997) Stem pull-through (courtship). Pull leafy branch of shrub or clump of grass stems through hand by rapid upward movement of forearm. Then immediately release stem, producing conspicuous sound, while seated and watching estrous female. Only two of oldest adult males, Musa and Ntologi, did this courtship. Exaggerated form of <pull through with hand> in eating context. One adult male (Fanana) tore off shoot of stem from shrub in courtship. Customary at Bossou (Whiten et al. 1999). (Type A) Category 8.

pull through with hand (PTH)
Upward or outward sweeping movement of thumb and index finger or palmar surface of hand along leafy branch or leafy herb stem. Common food-processing technique for leaves, vine, herbs or blossoms (e.g. *Erythrina abyssinica*). Leaf-stripping technique also used to make tool for fishing carpenter ants. See <fish>. Gombe chimpanzees use in ant dipping

(McGrew 1974). Seen in bonobos (Kano unpublished). (Type A) Category 1–2. See video. See also videos "eat leaf" and "store."

pull through with mouth (PTM)

Hold twig with hand and pull it between lips or teeth. Common food-processing technique for soft leaves and blossoms of trees, vines, and herbs. Used to make tool for fishing carpenter ants. (Type A) Category 3–7. See video.

pull with mouth (PWM)

Mouth and pull body part of another, in social play. If body dragged extensively, is <drag other by mouth>. (Type A) Category 3–7. See video.

punch

See <hit>.

puncture (PUN)

Dig underground termite nest with stout stick. Stand bipedal, push stout stick with foot, while holding stick with hands. Customary at Goualougo (Sanz et al. 2004) and recorded at Ndoki (Suzuki et al. 1995). Absent at Mahale. See video (Goualougo).

push (PUS)

Extend arm or foot to make forceful contact in aggression, grooming, infantile courtship, social play, and play solicitation. To groom another, push part of back (e.g., shoulder), so that groomee turns torso. Groomer stimulates groomee to change posture after extensive grooming of body-part. Mother pushes infant to descend from her back to ground. Mother gives infant's back or rump swift push with hand or foot to make it walk ahead. Juvenile male sometimes pushes back of estrous female to solicit copulation. Kano's (1998) and Pika's (2007) Push in bonobos. (Type A) Category 1. See video.

push ahead (PUA)

Mother drops off older infant from back or belly and encourages it to go ahead by pushing or pulling so as to travel comfortably and secure infant's security. Goodall (1989): "A mother pushes her infant ahead of her during weaning to prevent dorsal travel. Often she does this with little slaps of her hand on its rump and this is apparently perceived by the infant as playful". Whether or not "playful slaps" is unclear at Mahale. Absent in bonobos of Wamba (Kano 1998). (Type C) Category 3–5. See video.

push and pull sponge (PPS)

Whiten et al. (1999) Sponge push-pull. Put leaf-sponge into hole of tree and pull it out with stick (Sugiyama 1997). At Mahale, two juveniles used stick to pick out leaf-sponge from tree hole (Matsusaka et al. 2006). Present also at Gombe and Taï. (Type B) Category 9.

push away (PAW)
Shove away infant with palmar side of hand forcibly, or back side of hand gently. Goodall (1989) Push: "Shove away from another individual with one or both hands…", and Goodall's (1989) Push away: "When a mother pushes her child away when it approaches to suckle or ride on her; also occurs during begging when adults are pushed away". Adult male shoves away older infant from mother when its approach hinders grooming. Kano's (1998) Push away for bonobos. (Type A) Category 1–2. See video.

push backward (PUB)
Push rock backward with hand while running forward in riverbed. Element of charging display of some adult male chimpanzees of M group. Not in bonobos (Kano unpublished). (Type A) Category 8. See video.

push down (PUD)
Adult male, Masudi, runs while pushing down short plants, such as grasses and shrubs one after another, in policing. (Type A) Category 8. See video.

push finger into mouth (PFM)
Nishida (1983a): "Insert one or more fingers/toes into the mouth of another individual." Goodall's (1989) Finger in mouth includes both <push finger into mouth> and <take finger in mouth>. Appeases or reassures another (usually, more dominant) individual, or calms excited adult male. Mother (Mija) uses to stop infant's threat-like vocalization. Not in bonobos, except when infants or juveniles beg for food (Kano 1998). (Type A) Category 2–6. See video.

push forward (PUF)
Juvenile female, Maggie, playfully and repeatedly pushes hard-shelled fruit forward on ground. Adult male, Alofu, repeatedly pushes stone forward, facing consorted female, Abi, to threaten that if she flees he will throw stone at her. (Type A) Category 8. See video.

push head into ventral (PHV)
Social play with "wrestling": Youngster pushes crown of head into playmate's chest or abdomen. Goodall's (1989) Butt. (Type A) Category 3–5. See video.

push leaf-pile (PSL)
Nishida and Wallauer (2003): Walk forward quadrupedally while pushing dry leaves with both hands (sometimes, with feet). Locomotory-object play done mostly by juveniles. Adults and adolescents use to threaten other. Cf. <drag dry leaves with feet> and <pull leaf-pile>. (Type B) Category 7. See video.

push object into (POH)
Stuff leaves, leaf sponges, or some other objects into tree hole. Occurs when drinking water by leaf sponge, inspecting inside of cavity, etc. Cf. <probe>. (Type A) Category 3. See video.

push peri-anogenital region with finger (PPA)
Knead peri-anogenital region with extended fingers to extract hardened feces, helping softer feces inside to be passed. (Type B) Category 1–8. See video.

push/pull swing (PUW)
Mother or play partner pushes and/or pulls back and forth with hand youngster or vertical woody vine to which it clings. (Type C) Category 1–8. See video.

push shoulders
See <grasp and push shoulders>.

push to shoulder (PSS)
Push object against shoulder with hand. Function unknown. (Type A) Category 3–8. See video.

put dorsal (PTD)
Nishida (1983a) and Goodall (1989) Take dorsal. When infant not on mother, she grasps infant's arm and brings to her shoulder to place onto her back. Cf. <scoop infant>, <put ventral>, and <solicit riding>. Also in bonobos (Kano 1998). (Type A) Category 2. See video.

put dorsal from ventral (PDV)
Goodall (1989) Take dorsal. Caretaker transfers infant from belly to back with scooping upward movement of hand. Cf. <put ventral from dorsal>. Also known for bonobos (Kano 1998). (Type A) Category 2. See video.

put face close
See <peer>.

put face to (PTF)
Put face to carcass, apparently enjoying feel of soft hair. Done by adolescent male, Cadmus. Rare behavior. (Type A) Category 9. See video.

P

put heel on back (PHB)
Adolescent male, Cadmus, put right heel on back of adult female, Totzy, in response to her pant-grunting to him. (Type A) Category 9. See video.

put mouth into water (PMW)
Immerse mouth in stream but without drinking. Water-contact play of juvenile chimpanzees in quadruped or lying posture. (Type A) Category 8. See video.

put rump to rump (PRR)
Goodall (1989) Rump-to-rump: "In response to a submissive present, the higher ranking chimp turns his rump to the other and they touch bottoms. Only seen in a few males." Very rare at Mahale and in few males and females. Cf. <rub genital>. Common in bonobos (Kano 1998). (Type A) Category 2–8. See video.

put ventral (PTV)

Goodall (1989) Take ventral: "Mother reaches to infant, grasps, and presses to a ventral position." Cf. <put dorsal>, <scoop infant> and <solicit riding>. Kano's (1998) Take ventral in bonobos. (Type A) Category 2–6. See video.

put ventral from dorsal (PVD)

Mother transfers infant from dorsal to ventral position, when entering dense bush or meeting possible danger, such as unknown humans. Cf. <put dorsal from ventral>. (Type A) Category 2–6. See video.

Q

quadrupedal hunch
See <hunch quadrupedal>.

quadrupedal jump
See <hop quadrupedal on spot> and <leap quadrupedal>.

quadrupedal run
See <run quadrupedal>.

quadrupedal transport
See <transport quadrupedal>.

quadrupedal walk
See <walk quadrupedal on knuckles>.

R

rain cover
See <cover from rain>.

rain dance
See <display, rain>.

raise (RAI)
Elevate limb while lying supine or sitting on ground. Also in bonobos (Kano unpublished). (Type B) Category 1. See video.

raise and hold leg (RAH)
Raise leg and hold with one or both hands while lying supine. Pattern of grooming solicitation and also in resting. In grooming and resting in bonobos (Kano unpublished). (Type A) Category 2. See video.

raise arm(s) bipedal (RAT)
Extend arm(s) and direct palm(s) to companion. Intention movement to embrace or pull individual overhead during play or stressful confrontation. In stress, usually done by grinning adult male. In play, usually done by youngster with play face. Also, used to grasp fruit in tree, or to climb tree. Infant does without anything overhead when requesting mother to allow access to nipple or to seek rescue. Signal for help probably originated from intention movement to seek dorsal ride on mother. (Type A) Category 1–6. See video. See also video "approach and withdraw."

raise arm quickly (RAQ)
Goodall (1989) Arm threat: "The arm (whole or forearm only) is raised in a quick jerky movement and the fingers flexed slightly. This is usually accompanied by head tipping and cough threat or waa bark." Plooij's (1984) arm raise (ARA). Cf. <shake arm>. Adolescent male, Cadmus, raises arm quickly while standing bipedally for courtship as well as intimidation. Also in bonobos of Wamba (Kano unpublished). (Type A) Category 2. See video.

raise arm slowly (RAS)
Extend arm overhead, seeking to solicit grooming. See also <solicit grooming>. Also in bonobos (Kano unpublished). (Type A) Category 2. See video.

T. Nishida et al., *Chimpanzee Behavior in the Wild: An Audio-Visual Encyclopedia*,
DOI 10.1007/978-4-431-53895-0_9, © Springer 2010

raise arm to hold branch (RAB)

Raise arm and grasp overhead branch or liana to solicit grooming. See also <solicit grooming>. Also in bonobos (Kano unpublished). (Type A) Category 2. See video.

raise arm with elbow bent (RAE)

Extend arm overhead with bent elbow, to elicit grooming. If accompanied by self-scratching of armpit may stimulate grooming of armpit. Invitation for grooming-hand-clasp often complied with by mirror-image gesture from other, leading to grooming-hand-clasp. Also used when mother solicits infant to ride on her belly or back at onset of travel. See also <solicit grooming>. Also in bonobos (Kano unpublished, Fruth et al. 2006). (Type A) Category 2. See video. See also video "solicit grooming"

raise leg while lying (RAL)

Lie supine or sideways and raise leg, soliciting grooming of thigh, scrotum, or leg. Probably same as Goodall's (1989) ventro-ventral present: "When the sitting subordinate faces the dominant, leans back, raises one foot and rotates thighs laterally". Also in bonobos (Kano unpublished). (Type A) Category 2. See video. See also video "groom"

raise other's arm (ROA)

Elevate another's arm. Groomer seeks to change groomee's posture, so as to change body part to be groomed, and for groomer's convenience. Also in bonobos (Kano unpublished). (Type A) Category 2. See video.

raise other's chin (RAJ)

Elevate or lift up another's lower jaw with hand, to groom exposed neck or jaw. Also in bonobos (Kano unpublished). (Type A) Category 3–7. See video. See also video "groom unilaterally"

raise other's leg (ROL)

Same function as <raise other's arm>. (Type A) Category 2. See video.

rake (RAK)

R

Rake or pull dry leaves or sand toward self with one or both hands, while lying, sitting or quadrupedal. Solo play of youngsters. Used also as solicitation of play. Plooij's (1984) Rake (RAK). See also <pull leaf-pile>. (Type A) Category 3–7. See video. See also video "sway woody vegetation."

rake-hunch

See <scratch dead leaves>.

rap (RAP)

Nishida (1997) Thump: Courtship pattern of males and females; sit and hit stem of shrub, branch, or ground with knuckles. Customary at Taï and Mahale. Habitual at Gombe. (Type A) Category 3–8.

rapid ohoh

See <pant-grunt>.

reach
See <extend>.

reach hand
See <extend hand>.

reach leg
See <extend leg>.

reach wrist toward (RWT)
Goodall (1989) Wrist towards. Reach flexed wrist and back of hand to another. Cf. <extend hand> which is limited to gesture with wrist and fingers extended. At Mahale, subordinate behavior by some adolescents and adult females when pant-grunting to adult males, or threatened by dominant females. Also when prohibiting youngster from approaching infant. (Type A) Category 3–6. See video.

reassurance suck
See <suck in reassurance>.

reassure (REA)
Response to submission or fearful expressions. Shown mostly by higher-ranking individuals. Goodall's (1989) Reassurance includes <pat>, <touch>, <embrace>, <kiss>, and <groom>. Behavioral patterns exist in bonobos, but not in reassurance (Kano 1998). (Type D) Category 1–3. See video.

rebuff play (RBP)
Ignore play solicitation from another. Cf. <solicit play>. Cf. Hayaki's (1985a) Paraplay, which is unsuccessful attempt to initiate play with another. (Type D) Category 1. See video.

reciprocal grooming
See <groom reciprocally>.

reconcile (RCL)
de Waal (1982, 1989) Reconciliation. Post-conflict behavioral patterns between combatants accompanying body contact, such as <approach>, <touch>, <embrace>, <mount>, <kiss with open mouth> and <groom>. Reported from Gombe (Goodall 1986), Mahale (Kutsukake and Castles 2004) and Taï (Wittig and Boesch 2003). Bonobos of Wamba reconcile by <approach>, <touch>, <mount>, <groom>, <rub genitals>, and <put rump to rump> (Kano unpublished). (Type D) Category 1. See video.

recover
See <retrieve infant>.

recruit support
See <solicit support>.

redirection of aggression

See <aggress, redirected>.

regulate direction (RGD)

Mother interferes in traveling of older infant by pulling or pushing until it changes direction of travel as she wants. Also, adult male forces estrous female to travel as he wants by threat and attack. See <herd>. Cf. <depart together>. (Type D) Category 1–3. See video.

reingest vomit (RIV)

Consume own vomit. Goodall (1989) Reingesting vomit: “A sick chimp may vomit into one hand, then eat the vomit”. Not seen in bonobos at Wamba (Kano 1998). (Type A) Category 3–5. See video.

reject (REJ)

Turn down partner’s request, e.g. to groom, care alloparentally, mate, suckle, play, transport, etc. (Type D) Category 1. See video.

reject infant (RJT)

Goodall (1989) Rejection of infant by mother: Maternal behavior “preventing her infant from (1) having access to the nipple, (2) climbing dorsal or ventral, (3) sharing her food or (4) sleeping in her bed”, when the infant is being weaned. Behavioral patterns include (1) “<cover nipple>, <fend>, <turn around>, <reject-move>, (2) <reject-move>,<shrug>, <push ahead>, <reject-sit>, (3) <turn around>, <fend>, <clamp elbow>, and <reject-move> and (4) <threaten>”. See also <wean>. (Type D) Category 1. See video.

reject-move (RJM)

Goodall (1989): “Mother avoids infant attempting to suckle, ride on her or beg food by moving away.” May occur in bonobos (Kano unpublished). (Type C) Category 2. See video.

reject-sit (RJS)

Goodall (1989): “Mother sits when the infant is riding on her back and leans slightly back, so the infant can no longer ride. She may sit down the moment the infant climbs onto her back, or during travel...” Clear example unknown for bonobos (Kano 1998). (Type C) Category 3–5. See video.

relaxed face (RLF)

Goodall (1989) Relaxed face. Mouth closed and eyes open in normal circumstances. Kano’s (1998) Relaxed face. (Type A) Category 1. See video.

relaxed open mouth display

See <play face>.

release infant to fall (RIF)

Juvenile grasping hand of infant playmate released it, and let infant fall to ground. Seen once. (Type A) Category 8–9. See video.

remove (RMV)
Divest object from self. Pick up with thumb and forefinger small objects, such as sticky seeds of grasses or beggar's lice from body surface, in grooming. Also in bonobos. (Type D) Category 3–7. See video.

remove lice (RML)
Groomer pulls from hair with hand and removes louse egg, or puts teeth directly to hair to pull off louse imago. See also <scrape>. (Type B) Category 1. See video.

remove objects from water surface (RMO)
When streams begin to dry up, remove dry leaves, stones, or other debris, to search for good drinking pool. With digging, important technique to obtain water, <pick up>. At least five chimpanzees at Mahale do this. Cf. <dig for water>. (Type C) Category 3–8. See video.

remove sand flea (RMF)
Prise open toe and toe nail with both hands and suck out sand flea eggs (Nishida 2002). (Type B) Category 3–8. See video.

rescue (RSC)
Mother or allomother rushes to scene of attack, retrieves youngster. (Type C) Category 1. See video.

resin-pound
See <pound resin>.

respond to baboon (RSB)
Upon encountering yellow baboons, bark at, chase, or charge them. Chimpanzees of Mahale recently began to prey on yellow baboons (Nakamura 1997, Nishie 2004). (Type D) Category 3–7. See video.

respond to dead animal (RSA)
Encounter intact dead bodies of red colobus monkey, red-tailed monkey, bushpig, bushbuck, aardvark, giant rat, leopard, etc. Approach and stare fixedly at carcasses of large-sized animals. Do not eat old carcass, but rarely eat fresh one, which they had not killed. Responses to fresh adult leopard carcass included <inspect>, <watch>, <sniff>, <slap>, <hit>, <bite>, <groom>, <lift and drop>, <drag>, <pull>, <throw at> (Nishida unpublished). See also <eat carcass>. (Type D) Category 3–8. See video.

respond to dead chimpanzee (RSD)
Mother carries dead infant for weeks to months (Nishida 1973; Teleki 1976; Matsuzawa 1997). Other chimpanzees <approach>, <stare fixedly>, <sniff> and <whisk fly with arms> from corpse. May react to dead adult group mate by emitting <wraa> calls (Hosaka et al. 2000). May beat, bite, mutilate or eat dead chimpanzee (Boesch 2009; Nishida 1998, unpublished). See <transport corpse of infant>. Cf. <abuse carcass>. (Type D) Category 3. See video.

respond to leopard (RSL)

React to leopard's wood-sawing call by emitting <wraa>, <bark> or <pant-hoot>. Youngsters approach adults for reassurance. Chimpanzees killed leopard cub at Mahale (Hiraiwa-Hasegawa et al. 1986) and Taï (Boesch 2009). (Type D) Category 7. See video. See Fig. 31.

Fig. 31 Respond to leopard: Adult male Bakali slapped leopard cub (By courtesy of Richard Byrne)

respond to lion (RLI)

React to lion's roaring at night by emitting "fear calls"(<wraa>) then pant-hoot. Also, chimpanzees shifted daytime range to higher altitudes (Tsukahara 1993). Chimpanzees escaped to tree canopy and emitted <wraa> calls when chased by lions in savanna woodland of Ugalla (Itani 1979) and Assirik (Tutin et al. 1981). (Type D) Category 7.

respond to neighboring unit group (RNG)

Upon hearing calls from neighboring group, chimpanzees respond by pant-hooting and charging displays. They embrace each other and listen. If party is few in members, they silently retreat to center of territory. (Type B) Category 3. See video.

respond to predator (RSP)

Chimpanzees stalked, attacked, killed and eaten by predators such as lions, leopards and humans. In west and central Africa, chimpanzees are objects of bushmeat trade (Peterson and Ammann 2003). At Mahale at least six chimpanzees were taken by lions (Tsukahara 1993; Inagaki and Tsukahara 1993), but no record of predation by leopard. At Taï several chimpanzees killed by leopard (Boesch 1991a). No record of predation on chimpanzees by bird of prey at Mahale. See <respond to lion> and <respond to leopard>. (Type D) Category 7.

rest (RES)

Goodall (1989) Rest. Relax and remain immobile, <sit>, <lean> or <lie>. Kano's (1998) Resting. (Type D) Category 1. See video.

restrain (RER)

Prevent partner from continuing ongoing behavior to its or partner's advantage. Cf. <interfere>. (Type D) Category 1. See video.

restrict

See <detain>.

retaliate (RTL)

Counter-attack, that is return attack to initial attacker. Goodall (1989): "Aggressive response directed toward the original aggressor by a chimp who has been threatened or attacked." In bonobos, retaliation occurs more often when females are attacked than when males are targeted (Kano 1998). Kano (1998) included as retaliation, dominant's response to challenging subordinate ("Punish attack"). Cf. <protest>. (Type D) Category 1. See video.

R

retreat (RET)

Walk backwards quadrupedally. Subordinate pant-grunts to dominant, and occasionally walks backwards fearfully in front of him. Adult male walks backwards, facing estrous or semi-estrous female, to confirm her following him. Also in bonobos. Cf. <creep>. (Type A) Category 2. See video.

retreat bipedal (RRB)

Walk backwards bipedally. Juvenile does rarely, in order to solicit play from partner following it. (Type A) Category 3–9. See video.

retrieve infant (RTR)

Nishida (1983a) Recover. Mother returns to infant left alone or with others, touches, grasps or transports it, in order to depart, groom, suckle, etc.

Occurs after infant emits distress calls, when danger threatens, or when mother wants to depart. Also in bonobos (Kano unpublished). (Type D) Category 1. See video.

return infant (RTI)

Alloparent returns infant to mother, when infant emits distress calls. Absent in bonobos of Wamba (Kano unpublished). (Type D) Category 3–6. See Fig. 32.

Fig. 32 Return infant: Adolescent female, Wakilufya, returned infant to its mother (T. Nishida)

reunite (REU)

Individuals meeting after separation show friendly (see <greet>) or aggressive behavior (see <show off>). (Type D) Category 1.

reverse (REV)
Turn and travel in opposite direction. (Type D) Category 1. See video.

ride bipedal (RDB)
Infant stands bipedal on mother's back, to reach branch overhead or just before leaping down to ground. (Type A) Category 3–7. See video.

ride clinging (RDC)
Goodall (1989): "Infant clings to mother's arm or leg as she travels." Also in bonobos (Kano unpublished). (Type A) Category 2. See video.

ride dangling (RDA)
Infant hangs below mother with one or both hands only. Kano's (1998) Dangling riding for bonobos. (Type A) Category 2. See video.

ride dangling and touch dry leaves (RDL)
One- to two year-old infant hangs below mother with hand and touches dry leaves on ground, apparently enjoying the sound produced. Solo play in dry season at Mahale. (Type B) Category 7–8. See video.

ride dorsal (RDD)
Infant approaches, puts hands on mother's buttocks, and climbs up to ride on her back, or occasionally, another individual. Goodall (1989) discriminates three infant dorsal positions: Jockey, Lying and Quadrupedal. Kano's (1998) Dorsal travel for bonobos. (Type B) Category 2. See video.

ride jockey (RDJ)
Goodall (1989): "When an infant sits on the back of another in a straight upright position. Its legs may be drawn up, with the feet on the other's back, or gripping around the other's side." Kano's (1998) "Dorsal travel: jockey style" for bonobos. (Type A) Category 2. See video.

ride onto
See <step on>.

ride prone (RDP)
Infant travels by lying prone on mother's back. Kano's (1998) "Dorsal travel: cling style" for bonobos. (Type A) Category 2. See video.

ride quadrupedal (RDQ)
Goodall (1989): "When an infant stands on the back of another." Kano's (1998) "Dorsal travel: quadrupedal style" for bonobos. (Type A) Category 2. See videos.

ride supine (RDS)
Infant travels by lying supine on mother's back, occasionally and briefly. (Type A) Category 3–8. See video.

ride ventral (RDV)
Goodall (1989) Ventral ride: "The infant is transported as he clings to the mother's belly, gripping hair between flexed fingers and toes...There are

following variations: Extended, Dangle, Arm or Leg-cling." Kano's (1998) Ventral riding for bonobos. (Type B) Category 2. See video.

ride ventral with limb extended (RVE)

Goodall (1989) Ventral riding-extended: Ventral infant grips mother's hair with four limbs extended, so its back almost brushes ground as she travels. Seen in bonobos (Kano unpublished). (Type A) Category 2. See video.

rinse (RIN)

Shake object sideways repeatedly in water. See <wash>. (Type A) Category 8–9.

rock back and forth (ROS)

Move torso with rhythmic, alternating forward and backward movements while sitting or standing bipedal, in threat. Also at onset of charging display. Seen also during <leap bipedal>. Plooij's (1984) Rock (ROC). (Type A) Category 3. See video.

rock side to side (ROC)

Goodall (1989) Rock: "Slight or vigorous side to side movements of the body when the chimp is sitting...Rocking occurs when a male is working up to a charging display..." At Mahale, rocking prelude to charging display and when nervous. Bonobo rocking occurs only in courtship, not as warming up for charging display (Kano 1998). (Type A) Category 2. See video.

roll (RLL)

Goodall (1989) Roll: "A chimpanzee may hit another and roll over him during aggression or play." Bonobos use in aggression (Kano unpublished). (Type A) Category 2. See video.

rooting

See <nuzzle>.

rotate fruit (RTF)

Juvenile male, Michio, lies supine on ground, extends legs upward and rotates big fruit such as *Voacanga lutescens* with his feet. Solo object play. (Type A) Category 9. See video.

R

rough and tumble play

See <play, rough and tumble>.

rub dorsum (RBD)

Lean against vertical tree trunk and rub wet shoulders or back against it, often holding arms upward, after heavy rain. May lie supine on rock, ground, grass, horizontal bough, or day-bed to rub against surface. Recorded at Gombe (Goodall 1968). Also in bonobos of Wamba (Kano unpublished). (Type A) Category 2–5. See video.

rub dorsum to conspecific (RBC)

Juvenile may lean against back of adult male to solicit play. Rare at Mahale. Idiosyncratic? (Type A) Category 9. See video.

rub genitals (RBG)
Female rubs genitals against another female. At Mahale rarely seen, and limited to pairs in which one participant is juvenile or adolescent. At Bossou more frequent and ventro-dorsal, and "put rump to" position seen (Zamma and Fujita 2004). Bonobo females often show species-specific "G-G (genito-genital)-rubbing" as reassurance (Kano 1992). (Type A) Category 8. See video.

rub genitals to substrate (RUG)
Female rubs genitals against substrate such as ground or tree trunk. (Type A) Category 1–9. See video.

rub hand or foot (RBH)
Corp et al. (2009) Hand-rubbing. Rub hand or foot against branch of tree, stalk of shrub, rock, or ground when eating juicy fruit of lemon or sticky fruit such as *Saba*. After touching feces or other dirty object, immediately rub soiled hand or foot against ground. Cf. <brush away from branch>. (Type B) Category 1–7. See video.

rub hand with hand (RHH)
Rub hands together briskly and repeatedly, e.g. after removal of ectoparasite from body. (Type A) Category 3–7. See video.

rub muzzle (RBM)
Corp et al. (2009) Muzzle-rub. Rub muzzle stained with fruit juice against shrub, grass stalk, slender vine, bark of tree, or stone, to wipe off mess. Differs from <wipe with detached object> because body-part put to "part of the environment", not tool to body. At Mahale, only one adult female did this, after eating seeds of *Parkia filicoidea* several times before 1990s. Became common in 1998; by 2004, 29 chimps of M group did so when eating juicy fruit of lemon or sticky fruit of *Saba comorensis* and *Voacanga lutescens*. Goodall (1989) omitted but occurs at Gombe (Corp et al. 2009). (Type A) Category 7. See video.

rub object to body (ROB)
Move quickly object such as leafy branch along body, back and forth. Adult male, Pim, rubbed large dead bird on nape and head while pressing it firmly. Juvenile female, Athena, lay on her back and rubbed her belly with smooth stone. (Type A) Category 8–9. See video.

rummage (RUM)
Search for target object by removing or separating other objects. (Type C) Category 1. See video.

rump to rump
See <put rump to rump>.

rump turn
See <present with limbs extended>.

run (RUN)

Rapid locomotory gait with which body momentarily off ground in each stride cycle. (Type B) Category 1. See video.

run away

See <flee>.

run bipedal (RUB)

Goodall (1989) Bipedal run. Chimpanzees run on feet only during charging display. Adult male, Masudi, once ran on feet over ground swarming with *Dorylus* ants, to avoid being bitten. Also in bonobos (Kano 1998). (Type A) Category 1. See video.

run quadrupedal (RUQ)

Goodall (1989) Quadrupedal run. (Type A) Category 2. See video.

run tripedal (RUT)

Run using feet and arm while other arm carries object. See <drag branch>. Kano's (1998) Tripedal run. (Type A) Category 2.

rush to embrace (RSE)

Frightened youngster dashes into arms of older individual, usually mother, often followed by sucking. (Type A) Category 2–3.

S

safari
See <consort>.

Salonga
See <Lui Kotale>.

sand
See <play with sand>.

sand flea
Jigger (*Tunga penetrans*). Common ectoparasite that often burrows under toenails and lays eggs, annoying human, chimpanzee and baboon.
See also <remove sand flea> and <medicate self>.

scan (SCN)
Look in all directions seeking something. Lost juvenile often does so seeking mother after climbing tree. Adults do when looking for others.
See also <search for conspecific>. Also in bonobos (Kano unpublished).
Cf. <monitor>. (Type D) Category 2.

scatter dry leaves (SCL)
Rake, shift to left or right, around, or throw forward dry leaves with hand. Infantile expression of dissatisfaction or protest, or solo play.
Cf. <scratch dry leaves>, which adults show as prelude to charging display.
(Type B) Category 3–7. See video.

scavenge
See <eat carcass>.

scoop algae (SCO)
Scoop algae from water using wand of vegetation. Customary at Bossou (Matsuzawa et al. 1996; Ohashi 2006), and at least one adult male at Odzala (Devos et al. 2002), but not elsewhere. Cf. <eat algae>. (Type B).
See video (Bossou).

scoop infant (SCI)

Goodall (1989): "When the mother is about to travel and the infant is on the ground near her rump she reaches back and pushes the child up onto her back with a backward and upward movement, using the palm of her hand and fingers." Cf. <put dorsal>, <put ventral> and <solicit riding>. Plooij's (1984) Scoop (SCO). Kano's (1998) Scoop for bonobos. (Type A) Category 2. See video.

scramble for food (SRF)

Compete for precious food such as meat by agonistic gestures and calls. (Type D) Category 1. See video.

scrape (SCP)

Scrape lice eggs through hair with incisors (Zamma 2002a). Also resins of trees such as *Terminalia mollis*. See also <eat resin>. (Type A) Category 3–7. See video.

scratch (SCR)

Rake nails of partly flexed fingers over surface. Goodall's (1989) Scratch. Plooij's (1984) Scratch (SCR). Seen as Self-scratch in bonobos (Kano 1998). (Type B) Category 1. See videos.

scratch aggressively (SCA)

Scratch another to inflict wound. Goodall's (1989) Aggressive scratch: "A chimpanzee may scratch another during a fight." Probably also in bonobos (Kano unpublished). (Type A) Category 1.

scratch dry leaves (SDL)

Scratch or rake with one or both hands (rarely also with foot) dead leaves piled on ground, as initiation of pant-hooting or charging display, as solicitation of grooming, as solicitation of play, or as courtship. Done by infants as solo play. Maybe same as Goodall's (1989) Scrub ("A component of charging displays. The chimp takes a bunch of vegetation, burlap nag etc., and moves his hand in a semi-circular motion from side to side. Often in preparation for a charging display") or Rake ("The sweeping of ground vegetation with straight arm movements while showing a quadrupedal hunch. Usually this occurs prior to a charging display"). Cf. <scatter dry leaves>. Plooij's (1984) Rake (RAK). Bonobos rake before dragging *Haumania* vine in display (Kano 1998). (Type A) Category 2–5. See video.

scratch self (SCE)

Goodall (1989) Self-scratch: "Nails drawn deliberately across skin. In self-grooming, scratching is directed against the direction of hair growth and is followed by grooming of the scratched part... " Goodall's (1989) Scratch self directly: "---during social grooming the groomee often scratches a body part that is then groomed by the partner." Occurs when frustrated and uneasy. Plooij's (1984) Self-scratch (SCS). Kano's (1998) Self-scratch for bonobos. (Type A) Category 1. See video.

scratch self directly
See <scratch self>.

scratch self signalling (SCD)
Goodall (1989) Distance scratch: "A grooming session sometimes initiated when one chimp sits, holds an overhead branch, and makes deliberate scratching movements while staring at the desired grooming partner who may be several meters away. Loud frustrated scratching may serve as a signal to a reluctant female who is being led away on a consortship, indicating that the patience of the male is wearing thin." Bonobos use at start of grooming (Kano unpublished). (Type C) Category 2. See video.

scratch socially (SCS)
Nishida (1983a): Scratch or Rub. Nakamura et al. (1999) Social scratch. Scratch vigorously skin of another's back or arm at beginning or middle of grooming bout, usually with one hand, making audible sound. Absent at Gombe as tradition (Nakamura et al. 1999) but seen in one male and two females in 2001 (Shimada 2002). Bonobos of Wamba occasionally scratch another with first and second fingers in grooming (Kano 1998). (Type B) Category 6. See video.

scratch socially, poke type (SSP)
When scratching another at Mahale, few rarely poke with straight fingers. Customary among chimpanzees of Ngogo (Nishida et al. 2004). (Type A) Category 8–9. See video (Ngogo).

scratch socially, stroke type (SSS)
When scratching another at Mahale, usually stroke gently with bent (occasionally, straight) fingers. (Type A) Category 7. See video.

scream (SRM)
Goodall (1989): "...high pitched and loud, almost always given in a series. Usually heard in contexts of aggression and general social excitement by highly stressed, fearful, frustrated, or excited individual..." See also <squeal in copulation> and <whimper-scream>. Kano (1998) divided bonobo scream calls into Scream and Screech. (Type A) Category 1. See video. See also video "monitor mother."

scream, food (SCF)
Adult male, Fanana, screams when eating prized food, such as ripe fruit of lemon, *Saba* or *Garcinia*, especially at beginning of feeding bout. Absent in Goodall (1989). Bonobo's Food peep (Okayasu 1991) or Food grunt (Kano 1998) may be homologous. (Type A) Category 8.

scrub
See <scratch dry leaves>.

scrub pelt (SCC)
Rub colobus skin against rock in streambed, apparently to soften it. Done by one adult male (Nishida 1993b). See <wash>. (Type A) Category 9.

search for conspecific (SEA)

Look for unseen others, in several ways: (1) sit and listen, (2) walk around, then return to original place, (3) sniff ground, in particular, at path intersections and footprints, or (4) climb tree and scan surroundings, frequently done by lost, whimpering juvenile (See also <scan>). Mother also looks for unweaned infant. Mature male looks for estrous female. Behavior includes: <scan>, <walk>, <run>, <climb>, <grin>, <listen>, <whimper>, <whimper-scream>, <scream>, <pant-hoot>, <sniff>, <look back>, etc. Absent in bonobos of Wamba (Kano unpublished). (Type D) Category 3–5. See video.

search for object (SEO)

Search for object, such as food, rock to throw, sleeping tree, etc. Includes: <scan>, <walk>, <climb>, <move>, <touch>, <pull>, <stare fixedly>, <sniff>, etc. (Type D) Category 1–3. See video.

seat-stick (SST)

Branch used to avoid sharp thorns when sitting on kapok tree (*Ceiba pentandula*). Reported from Tenkere (Alp 1997) and Goualougo (Sanz and Morgan 2007). Absent at Mahale. See also <stepping-stick>. (Type B)

seat vegetation

See <make leaf cushion>.

seed dispersal

See <disperse seed>.

self-clasp

See <clasp self>.

self-groom

See <groom self>.

self-inspect

See <inspect self>.

self-medication

See <medicate self>.

S

self play

See <play solo>.

self-scratch

See <scratch self>.

self-slap

See <slap self>.

self-suckle

See <suck self>.

self-tickle
See <tickle self with object>.

semen
Whitish rubbery liquid containing sperm ejected by penis. See <ejaculate>. See Fig. 33.

Fig. 33 Semen: Semen drips from female's bottom after copulation (T. Nishida)

Semliki
Study area of *Pan troglodytes schweinfurthii* within Toro-Semliki Wildlife Reserve, Uganda (0°50′–1°05′N, 30°20′–30°35′E). Research organized by K. Hunt from 1996 to present.

senescence
Symptoms of old age are presbyopia, cataract, greying hair on head, bent and balding back, worn teeth, loss of teeth, lack of estrus (in oldest females), slow movement, avoidance of climbing tall trees, reduced range, etc. See video.

separate (SEP)
Party splits into two or more smaller parties or subgroups. (Type D) Category 1. See video.

separating intervention
See <intervene to separate>.

sexual skin
Anogenital region of females. See <swelling of sexual skin>. See Fig. 34.

Fig. 34 Sexual skin: Sexual skin often bitten by adult males, so deforming shape (T. Nishida)

shake arm, abduct (SHA)
Sit and shake (abduct) horizontally-held arm towards another, in mild threat, often accompanied by <cough bark>. Cf. <raise arm quickly> and <hit toward>. (Type A) Category 3–5. See video.

shake arm, adduct (SHH)
Sit or stand bipedal and shake (adduct) horizontally-held arm at another, to come closer or solicit play. Probably originated as intention movement of <grab>, <grasp> or <hug>. Juvenile male, Xmas, once shook hand above water's surface while looking down at his reflection (Cf. <nod quadrupedal>). Sometimes in threat, if shaken quickly. (Type A) Category 7. See video.

shake back and forth
See <sway woody vegetation>.

shake back and forth with four limbs (SBF)
Cling to and shake tree trunk forward and backward using pushing force of feet and pulling force of arms. Threatening gesture. Infants and young juveniles do as solo play. (Type A) Category 3–7. See video.

shake body
See <shiver>.

shake branch (SHB)
Goodall (1989) Branch shake: "A branch is shaken with quick jerky movements of the arm, slight or vigorous. The chimp may be sitting or

standing..." Plooij's (1984) Branch (BRN). Nishida's (1997) Branch-shake: Courtship pattern of males and females: Sit while shaking horizontally (side to side or back and force) or vertically stem of shrub, branch of tree, or woody vine. Branch-shaking signals request to another to come closer or to depart together. Contexts include courtship, continuation of consortship (as herding technique), enticement to travel together, threat, and solicitation of play. Cf. <sway woody vegetation> in which chimp bipedally shakes tree trunk, etc. back and force (not side to side) forcibly with both arms. Cf. also <flail long object>. Kano's (1998) Branch-shaking and Ingmanson's (1996) Branch waving for bonobos of Wamba. Hohmann and Fruth's (2003) Branch shake for bonobos of Lomako. (Type B) Category 2. See video.

shake branch up and down with feet (SHU)
Stand on horizontal branch, woody vine, or fallen log on ground, and shake it up and down with one or both feet, in threat display or solo play. Weanling does as part of temper tantrum, namely, reaction to lack of response of mother or another, as attention-getting. Reported only at Mahale. (Type A) Category 3–6. See video.

shake detached branch (SHD)
Hold detached, leafy branch in hand or foot and shaken horizontally in courtship, done by several juvenile males. (Type A) Category 8.

shake face side to side (SHC)
Amuse self by watching reflection on water's surface. Juvenile male, Xmas, vigorously shook face side to side. Cf. <nod quadrupedal>. (Type A) Category 8–9. See video.

shake face with object in mouth (SFS)
Hold object such as leaf-sponge in mouth and vigorously shake face side to side in stream. Done as solo play by adolescent male, Cadmus, only. (Type B) Category 9. See video.

shake hand
See <shake arm, abduct> or <shake arm, adduct>.

shake hand side to side quickly (SHQ)
Shake hand from side to side quickly, while pant-grunting. Conflict behavior shown by subordinate chimpanzee to dominant male. (Type A) Category 7. See video.

shake hands
See <grasp hand>.

shake head (SHE)

Shake vigorously head side to side while trying to spray about water or sand atop head. Immatures shake head during play, such as <play walk>. Cf. <tilt head> and <nod quadrupedal>. (Type A) Category 3–5. See video.

shake long object irregularly (SHL)

Adolescent male, Cadmus, sits and irregularly shakes long dry grass, both vertically and horizontally, as threat to human observer. (Type A) Category 9. See video.

shake object up and down (SHT)

Stand and shake up and down bush or dead tree by hand(s), in threat or social play. (Type A) Category 3–6. See video.

shake off (SHF)

While seated or standing, forcefully shake off play partner from body contact. Throw down opponent by hand in wrestling as in "sumo". (Type A) Category 3–6. See video.

shake other (SKN)

Old female, Calliope, grasped knee of Fanana (alpha male) and shook it back and force in greeting. Adult female, Xtina, grasped shoulder of another adult female, Ruby, and shook it back and forth when agitated. Adolescent male, Cadmus, grasped and shook juvenile male's heel as play. Idiosyncratic? (Type A) Category 9. See video.

shake penis (SHP)

Adult male grasps and shakes vertically another's penis, while sometimes panting. Rare at Mahale. Function as reassurance? Cf. <fumble penis>. (Type A) Category 8. See video.

shake rock up and down with feet (SHR)

Stand quadrupedal on rock and bounce up and down on it with feet. Solo play pattern of water-contact. (Type A) Category 7–8. See video.

shake rump (SRP)

Mother shakes lower back from side to side, causing infant to descend to ground. See <drop infant>. (Type A) Category 3–5. See video.

shake wet arm to catch termite (SCT)

Catch winged termites by shaking arms wet with rain. Technique employed for termites of *Pseudacanthotermes* (Kiyono 2008). (Type A) Category 8–9.

share bed (SHN)

Mother (or alloparent) allows infant offspring (or adopted orphan) to sleep in same bed. Cf. <allow>. (Type D) Category 1. See Fig. 35.

Fig. 35 Share bed: (T. Nishida)

share food (SFO)

Allow another to take food under control of owner (e.g. in mouth or hand, on lap, or in proximity). Same as Feistner and McGrew's (1989) Transfer of food. Most common between mothers and infants. Meat is most common food to be shared among adults. Meat sharing among unrelated adults occurs at Gombe (Teleki 1973), Mahale (Nishida et al. 1992), Taï (Boesch and Boesch-Achermann 2000), and Ngogo (Mitani and Watts 2001). Fruit sharing occurs at Gombe (McGrew 1975), Mahale (Nakamura and Itoh 2001), Budongo (Bethell et al. 2000; Slocombe and Newton-Fisher 2005), Bossou (Ohashi 2007; Hockings et al. 2007), Ngogo (Nishida, unpublished) and Taï (Nishida, unpublished). Large fruits such as *Treculia africana* and papaya often shared. Sugar cane sharing occurs at Mahale (Nishida 1970). "Food share" (Kano 1998) occurs in bonobos at Wamba and Lomako (Hohmann and Fruth 1993). (Type D) Category 1. See video.

shelter from rain (SLR)

Take cover from rain e.g., by going under canopy of big tree. Includes <walk> and <sit>. Cf. <cover from rain>. (Type D) Category 1–6.

shiver (SHV)

Chimps shiver when cold or wet. (Type A) Category 1.

show off (SHO)

Display by dominant individual in aggression between individuals meeting after long separation. Includes <bristle>, <charge>, <display, charging>, <hunch>, <slap>, <stamp>, <splash water>, <swagger bipedal>, <attack>, etc. In reunion, bonobos show <display, charging> in aggression (Kano unpublished). Cf. <display as contest>. (Type D) Category 2.

S

shrub-bend
See <bend shrub>.

shrug (SRG)
Goodall (1989): “Mother raises one shoulder slightly and lowers the other. Signals the infant to climb off its dorsal position. Gesture occurs during weaning and is a sign of rejection.” Plooij’s (1984) Shrug (SHR). Common at Mahale. See also <drop infant>. Cf. <lower head and shoulder>. Absent in bonobos of Wamba (Kano 1998). (Type A) Category 3–5. See video.

sickness
Symptoms of sickness include <cough>, <diarrhea>, <lethargy>, <lie>, <pick ear>, <pick nose>, <sneeze>, <sniffle>, <vomit>, runny nose, unstable walking, etc. See video.

silent bared-teeth display
See <grin>.

sit (SIT)
Plooij (1984) SIT. Goodall (1989): “Rump on ground, branch, etc., body more or less vertical.” Arms sometimes crossed. Kano’s (1998) Sit. (Type A) Category 1. See videos.

sit and lean (SIL)
Sit and extend both arms backwards and lean on them. Absent in bonobos. (Type A) Category 1–3.

sit and turn back (SIB)
Solicit grooming by approaching another, then turning back to potential groomer. See also Solicit grooming. Also in bonobos (Kano unpublished). (Type C) Category 2. See video.

sit behind another (SIA)
Change position to groom back of companion. Also in bonobos (Kano unpublished). (Type C) Category 2. See video.

sit face to face (SIF)
Shift posture to sit *en face*, in social grooming or seeking to receive food. Also in bonobos (Kano unpublished). (Type C) Category 1–2. See video.

S

sit on (SIO)
In social play, one player (often bigger one such as mother) sits on other player (such as juvenile offspring). Cf. <step on> in which smaller playmate ‘rides’ on larger one lying supine. Cf. also <trample>. (Type B) Category 3–8. See video.

sit prone (SIP)
Slouch or bend forward, extend arms, and put palms on ground, often when being groomed. Also in bonobos (Kano unpublished). (Type A) Category 2. See video.

sit sideways (SIS)

Shift posture to sit beside another, to solicit grooming. See also <solicit grooming>. Also in bonobos (Kano unpublished). (Type C) Category 2. See video.

sitting hunch

See <hunch and sit>.

slap (SLA)

Plooij (1984) SLA. Goodall (1989): "A hitting down gesture when the palmar surface of the hand is brought into contact with the objective...". Done with one or both hands. Context includes aggression, charging display, drumming, social play, solicitation of courtship and play, etc. Bonobos slap in aggression, charging display, drumming and social play (Kano 1998). (Type B) Category 1. See video.

slap belly

See <slap self>.

slap branch (SBR)

Whiten et al. (1999) Branch slap. Slap branch with palmar surface of one hand for getting attention. Few adult males of M group do in courtship display. Customary at Bossou, Taï and Budongo. Branch slap described for Lomako bonobos (Hohmann and Fruth 2003) may differ from chimpanzee pattern. Cf. <rap>. (Type A) Category 8.

slap buttress or tree trunk (SLB)

Strike tree with hand or alternate hands during charging display. Goodall's (1989) Drum. See also <drum> and <slap wall>. Included in Kano's (1998) Drum for bonobos. (Type A) Category 2. See video.

slap ground (SLG)

Strike substrate with hand or alternate hands during pant-hooting, threatening, charging display, fighting, and play. Slap ground and pant-hoot in response to pant-hoot calls from distance. Recorded from Beni (Kortlandt 1964). Bonobos slap ground while staring at opponent (Kano 1998), so context more narrow. (Type A) Category 2. See video. See also video "enter hole."

slap in invitation (SLI)

Nishida (1983a) Invitation slap. Repeatedly strike rock or ground with one or both hands, to entice youngster into play. Elements include: <slap ground>, <play face>, <stand quadrupedal>. Adolescent male, Cadmus, repeatedly slapped hand of adult female lying on ground with back of hand softly while showing play face. Absent in bonobos of Wamba. (Type D) Category 3–5. See video.

slap other (SLO)

Done during play, play invitation, charging display or fighting. Juvenile male, Xmas, slapped back of adult male while grooming him. Pika's (2007) Slap for bonobos. (Type A) Category 1. See video.

slap self (SLC)

Nishida et al. (2009) Slap belly. Slap own belly, chest, or thigh. Five-year-old juvenile male, Cadmus, struck belly with right hand to make drum-like sounds, while hanging from branch by left arm, even when alone. Solo play or response to approach of human observer. Other juveniles, Xmas and Michio, later began regularly to slap belly or chest in intimidation display and solicitation of play, while hanging in tree, standing bipedal or walking bipedal on ground. Male infant, Caesar, slapped belly and thigh in self play. Absent in bonobos of Wamba (Kano unpublished). (Type A) Category 8. See videos.

slap-stamp (SLS)

Goodall (1989): "Slapping with hands and stamping with feet together." Plooij's (1984) Slap stamp (SLS). Element of charging display. Absent in bonobos of Wamba (Kano 1998). (Type A) Category 3–5. See video.

slap wall (SLW)

Slap wall of metal house with one or both hands or alternate hands, while standing bipedal or quadrupedal. Done regularly by M group adult males at Mahale (Nishida 2003b) and by Kasakela males at Gombe. See also <slap buttress or tree trunk> and <drum>. (Type B) Category 5. See video. See also video "slap buttress or tree trunk."

sleep (SLP)

Lie or sit immobile with closed eyes, unconscious. (Type B) Category 1. See video.

slide down boulder (SDW)

Youngsters occasionally slide supine or prone down large slanted boulder. (Type B) Category 3–8. See video.

slide down vertically (SLD)

Goodall (1989) Fireman slide: "A chimp may slide down a vertical or diagonal pole or tree in an upright or crouched position. All four limbs are in contact with the pole or tree." Kano's (1998) Vertical slide. (Type A) Category 1. See video.

sloth position

See <hang in sloth position>.

sloth walk

See <walk in sloth position>.

smack lip (SML)

Goodall (1989) Lip smack: "Occurs during grooming. Mouth slightly opened and closed rhythmically." Plooij's (1984) LIP. Also at Taï (Boesch 2009). See also <clack teeth> and <sputter>. Similar in bonobos, but unknown if homologous to <smack lip> (Kano 1998). (Type A) Category 2–3. See video.

smell

See <sniff>.

snatch (SNA)

Take suddenly and forcibly object such as food from another, against resistance. Adult males snatch food, especially meat, from others of any age-sex class, mothers from immature offspring, and juvenile sons from mothers. Juvenile may snatch branch from playmate. In bonobos, mostly done by juveniles (Kano 1998). (Type C) Category 1. See video.

sneer (SNR)

Goodall (1989): "The upper lip is retracted, sometimes on one side more than the other, to expose part of the upper teeth. In the wild the context is associated with fear of a human or a human associated situation. Only a few chimps sneer." Plooij's (1984) Sneer (SNE). At Mahale few individuals sneer. Cf. <grin>. (Type A) Category 8–9. See video.

sneeze (SNZ)

Goodall (1989): "Chimps sneeze in the same way as humans." See also <probe nasal passage>. Kano's (1998) Sneeze for bonobos. See <sickness>. (Type A) Category 1. See video.

sniff (SFF)

Goodall (1989) Sniff. Put nose close to something, such as substrate, vegetation, tree trunk, fruit, food wadge, feces, urine etc. apparently to obtain information about which chimpanzee or what kind of animal was there. Sniff ground after left behind by companions, or when arriving at place where strange chimpanzees were before. Males sniffed more frequently than females in sexual and social situations, while females did so more often during feeding and self-checking (Matsumoto-Oda et al. 2006). Also reported from Taï (Boesch 2009). Bonobo's sniffing predominantly by juveniles (Kano 1998). (Type A) Category 1. See video.

sniff finger (SFG)

Plooij (1984) SFI. Goodall (1989): "One finger, usually index finger, is carefully smelled after being used to investigate some object. The most usual context is during inspection of the genital area... of a female." See <inspect genitals>. Juvenile bonobos do so (Kano 1998). (Type A) Category 1. See video.

S

sniff fruit (SFI)

Hold and sniff large hard-shelled fruits such as *Saba comorensis* and *Voacanga lutescens* to judge ripeness. (Type A) Category 1–3. See video.

sniff genital

See <inspect genitals>.

sniff mouth (SNO)

Sit or stand quadrupedally very close to another individual and sniff its mouth to obtain odor of food being eaten. Occurs especially for food new to immature individual. From behavior pattern only, difficult to discriminate from <peer> when peerer does not sniff. (Type A) Category 1–7. See video.

sniff with tool (SFT)

Goodall (1989) Investigation probe: "A stick or twig is used by a chimp for touching some object..." See also <probe>. Absent in bonobos of Wamba (Kano 1998). (Type A) Category 3–5. See video.

sniffle (SFL)

Audible sound of runny nose. See <eat nasal mucus> and <sickness>. (Type A) Category 1.

snub (SNB)

Rebuff by inaction. Apparently deliberate refusal to look at conspecific who approaches closely to seek contact. At Mahale, alpha male, Ntologi, did not look at beta male, Nsaba, even if Nsaba continuously followed Ntologi apparently seeking to groom him. Apparently signals intimidation that alpha male does not wish to interact with beta. Cf. <ignore>. (Type C) Category 1–8. See video.

social play

See <play socially>.

social scratch

See <scratch socially>.

soil-eating

See <eat termite soil>.

solicit companion (SOF)

When about to depart, adult encourages continued proximity from another resting nearby or about to leave in another direction by <look back>, <look through thighs>, <touch>, <pull>, <shake branch>, etc. Cf. <depart together> and <herd>. (Type D) Category 1–3. See video.

solicit copulation (SOC)

Courtship. Gesture and posture directed to estrous female by male or vice versa that normally leads to copulation. As male behavior toward estrous female, includes <clip leaf>, <bend shrub>, <pull through in courtship>, <swagger bipedal>, <hop quadrupedal on spot>, <hunch and sit>, <shake branch>, <open thighs>, <extend hand, palm downward>, <erect penis>, <bristle>, <gaze> repeatedly, <club ground>, <approach>, etc. Female solicits by <gaze>, <glance>, <clip leaf>, <approach>, <follow>, <dance bipedal> and <crouch>. <Clip leaf>, <bend shrub>, <dance bipedal>, etc. unique to Mahale (Nishida 1997). Goodall's (1989) Courtship includes "display brachiation". In bonobos of Wamba, <hunch bipedal>, <shake branch>, <rock>, <open thighs>, and <hunch and sit> are courtship displays (Kano 1998). (Type D) Category 2. See video. See also videos "aggress in sexual frustration" and "copulate dorso-ventral".

S

solicit grooming (SOG)

van Hooff (1973) Groom-present. Goodall's (1989) Present for (solicit) grooming: "The rump, back, or bowed head are the most usual body parts presented to a chosen grooming partner at the start of a session..." <Scratch self>, <raise arm>, <sit and turn back>, <sit sideways>, <raise leg>, <raise arm to hold branch>, <raise arm slowly>, <lie with back to another>, etc. also used. Alter position of limbs, head or body-part to facilitate continuation of grooming. Self-scratching most important of four patterns of solicitation for grooming in bonobos of Wamba (Kano 1998). (Type D) Category 2. See video.

solicit grooming turn (SOT)

Signal requesting partner to take turn in social grooming: (1) stop grooming and release partner's body, (2) stop grooming and scratch or groom self, (3) stop grooming and present back or other body-part to partner. Bonobos show similar behavior (Kano 1998). (Type D) Category 2. See video.

solicit play (SOP)

Invite another to play. Includes: <approach> with <play face>, <flee>, <extend hand>, <hit> lightly, <hit and run>, <hold object in mouth>, <lie supine>, <lie prone>, <slap in invitation>, <stamp in invitation>, <somersault>, <play face>, <play walk>, <throw branch>, <tickle>, <tilt head>, <watch>, and many others. Includes Goodall's (1989) Try play: "Unsuccessful attempt to initiate play with another" and Hayaki's (1985a) Paraplay. Bonobos invite play by <approach with play face>, <hit> lightly or <pull> the hand of partner (Kano 1998). (Type D) Category 1–2. See videos.

solicit play with object (SOO)

Invite play by holding plant part in mouth, hand, foot, or groin pocket. Customary at Gombe, Mahale, Kibale, Budongo, and Taï. Present at Bossou, Assirik, and Goualougo (Sanz and Morgan 2007). Also, includes <club ground>, <flail long object>, <throw>, etc. (Type B) Category 3. See video.

solicit play with object in mouth (SOM)

Whiten et al. (1999) Play start: Invite play by holding stem or other object in mouth. Probably chimpanzee universal. (Type A) Category 3. See video.

solicit reassurance contact (SOA)

Aggressed-against or frightened infant runs to mother, whimpers with extended hand or throws temper tantrum. Aggressed-against adult seeks contact with third party or with aggressor by extending hand, etc. Goodall (1989) mentioned but did not name pattern. Kano's (1998) Request for appeasement. Aggressed-against bonobo female pesters aggressor female and persistently peers at her, so that aggressor invites victim to G-G rub (Kano 1992). (Type B) Category 2. See video.

solicit riding (SOR)

Mother or allomother invites infant to ride on back or belly by <extend hand>,<hoo>, <look back>, <raise arm with elbow bent>, <touch> etc.

Cf. <depart together>, <put dorsal>, <put ventral> and <scoop infant>. (Type B) Category 1–3. See video.

solicit support (SOS)

Request support from third party when threatened or attacked. Includes: <search for conspecific>, <approach>, <extend arm>, <scream>, <touch>, <mount>, <thrust>, <kiss>, <embrace>, <press teeth against back>, etc. At Wamba, bonobos solicit only mother for support (Kano unpublished). (Type D) Category 1. See video.

solitary play

See <play solo>.

solo play

See <play solo>.

somersault (SMT)

Goodall (1989): Roll "head over heels". Plooij's (1984) Somersault (SUM). Usually repeated many times. Accompanies social play such as <circle> and <wrestle>. Also used as solicitation of play. Mature and immature chimpanzees do. At Mahale, 3–4 year old infants often do in solo play when large group travels long distance. See also <headstand>. Absent in bonobos of Wamba, but seems to occur in captivity (Pika 2007). (Type B) Category 1. See video.

somersault, backward (SMB)

Somersault backwards. Context of occurrence is similar to <somersault, forward> and <somersault, side>. (Type A) Category 1–5. See video.

somersault, backward with dry leaves (SBL)

Collect dry leaves on belly and somersault backward with them; while rolling, leaves scatter little by little. Juvenile males, such as Xmas and Primus, show in solo play. (Type A) Category 8. See video.

somersault, forward (SMF)

Usual somersault in forward direction. Often used as solicitation of play and during play. <Circle quadrupedal> usually accompanied by this type of somersault. Also done as solo play. (Type A) Category 1. See video.

S

somersault, side (SMS)

Somersault sideways. Used as solicitation of play and during social play, but also done as solo play. (Type A) Category 1–5. See video.

spar (SPA)

At start of fight, both parties stand bipedally and flail (without contact) one or both arms in hitting movements, with or without screaming. Also in play. Bonobo juveniles hit each other in play (Kano 1998). (Type A) Category 2.

spit juice (SPJ)

Forcefully expel juice of fruit such as unripe *Saba comorensis*, buds such as *Milicia excelsa*, or young leaves such as *Brachystegia spiciformis*. (Type A) Category 3–7. See video.

spit seed (SPT)
Spit large seeds such as *Garcinia huillensis*, *Voacanga lutescens*, *Syzigium guineense*, etc. at site where eaten. Also in bonobos. (Type A) Category 1. See video.

spit water (SPI)
Infant or juvenile occasionally spits water during drinking session. Solo play? (Type A) Category 1. See video.

splash water (SPW)
Solo play: Youngster splashes water with palm or fingers of hand, foot, leg, or mouth. Sprinkle water onto playmate. See also <play with water>. Cf. <stir water>. (Type A) Category 7–8. See video. See also video "play with water".

sponge
See <leaf-sponge>.

sponge push-pull
See <push and pull sponge>.

spring (SPR)
Infant rushes and jumps onto mother's buttocks to catch up with her rapid movement. (Type A) Category 2–3. See video.

sputter (SPU)
Nishida et al. (2004): Sound uttered by many chimpanzees of Ngogo but only few of Mahale when grooming another and finding something interesting. Usually, done in same context as <clack teeth> and <smack lip>. While sputtering they do not show rhythmic lip movements. Sounds as if chimpanzees force air through lips, or force saliva through teeth. Present at Bossou (Nakamura and Nishida 2006). (Type A) Category 8–9. See video (Ngogo).

squash ectoparasite on arm (SQE)
Whiten et al. (1999) Index-hit: Squash ectoparasite on arm. Customary only at Taï. See <squash ectoparasite on palm>. Absent at Mahale. (Type A)

squash ectoparasite on palm (SQP)
Nakamura and Nishida (2006) Squash ectoparasite on palm. Customary only at Bossou. Cf. <squash ectoparasite on arm>. Absent at Mahale. (Type A)

S

squash leaf (SQA)
Whiten et al. Leaf-squash: Squash ectoparasite on leaf. Mahale chimpanzees take louse from body, place it on leaf, fold leaf, and squash it with thumb and fingers (Zamma 2002a). Habitual at Gombe. See also <groom leaf>. (Type B) Category 4–5. See video.

squat (SQT)
Hunt et al. (1996): "The body weight is borne solely by the feet/foot, both hip and knee are strongly flexed." Also in bonobos. Cf. <sit>. (Type A) Category 1.

squat bob

See <bob>.

squeal in copulation (SQU)

High-pitched, scream-like sound accompanied by grin, emitted by estrous female at final stage of copulation. Goodall's (1989) Copulation scream (previously termed squeal). Bonobos emit similar call (Kano 1998). (Type A) Category 2. See video.

squeeze (SQZ)

Compress contents of mouth by pushing chin upwards. Hand or arms pressed against chin to facilitate wadging. See also <wadge>. Also in bonobos. (Type A) Category 2. See video.

staccato call (STC)

Newborn emits call in response to mother's sudden movement, approach by others, noises in surroundings, etc. Plooij's (1984) Staccato (STC). (Type A) Category 3. See video.

stalk (STK)

Approach stealthily to prey or other chimpanzees, especially neighbors. Absent in bonobos. (Type D) Category 3.

stamp (STA)

Goodall (1989): "Forceful downward kick of one foot, or alternate feet. (with sole making contact)." In charging displays, "The chimp stamps his feet one after the other as he runs along. One foot is often stamped harder than the other. A standing chimp may stamp alternate feet...A sitting chimp may stamp several times with one foot." van Hooff's (1973) Stamp. Plooij's (1984) Stamp (STA). Done also during social play, in solicitation of play and courtship. Cf. <kick>. Bonobos stamp while running, but not while sitting (Kano 1998). Pika's (2007) Stomp for captive bonobos. (Type B) Category 1–2. See video.

stamp bipedal (STB)

Stamp sole on ground, buttress, tree trunk, wall of house, or another while upright. Component of charging display. Adolescent males stamp with alternate feet on ground in courtship. Bonobos of Wamba stamp bipedal in courtship, but not in charging display. (Type A) Category 2–3. See video.

S

stamp in invitation (STI)

Nishida (1983a) Invitation stamp. Mainly youngsters, lightly stamp on ground while watching potential playmate, inviting social play or alloparental care. Bipedal or quadrupedal. Absent in bonobos of Wamba. (Type A) Category 3–7. See video.

stamp other (STS)

Goodall (1989): "Stamping on a victim is a major component of within-species fighting, and can result in severe bruising and contusions." Also stamp softly on playmate repeatedly in social play. Also in bonobos. (Type A) Category 1–2. See video.

stamp quadrupedal (STQ)
Stamp quadrupedal on ground, rock, buttress, tree trunk, wall of house, another, or carcass of animal. Component of charging display, etc. Juveniles stamp on rock or lying adult in play. Absent in bonobos? (Type A) Category 3. See video. See also video "pant-hoot."

stamp trot (STT)
van Hooff (1973): "The animal walks or trots along while stamping heavily with its hind feet on the ground. Often this is done in a rhythmical manner in that one foot is placed down gently and the other with force. The head is mostly kept rather low and tucked back between the shoulders which are pulled upwards. The head may make low amplitude vertical rocking movements ..." Kano's (1998) Stamping run and Pika's (2007) Stomp for bonobos. (Type A) Category 2. See video.

stamp water (SMW)
Juvenile female, Jiddah, while sitting, repeatedly stamped on small pool of water, as part of water play in stream. Adolescent male, Cadmus, bent branches onto water and stamped repeatedly on them as solo play. (Type B) Category 8–9. See video.

stand bipedal (SDB)
Plooij (1984) STB. Stand upright. Stance assumed when watching distant object or taking food from tall shrubs such as *Ficus urceolaris*. Mother may drop infant by almost standing bipedal. Cf. <walk bipedal>. Bipedal posture occurs in courtship, mixed with other elements, such as <shake branch>, <thrust>, <leap>, and <stamp>. Common in bonobos. (Type A) Category 1. See video.

stand quadrupedal (SDQ)
Stop walking and scan environment or wait for another, etc. Plooij's (1984) STQ. Common in bonobos. (Type A) Category 2. See video.

stand quadrupedal heel up (SQH)
While waiting for companion, or monitoring environment, stand still with heel up and toes on ground. Kano's (1998) Waiting posture for bonobos. (Type A) Category 2. See video.

S

stand with head down, bottom up (SWH)
Nishida (1983a): Head down hips up: Postural pattern with tickling by mouth. "The mother or caretaker stands on all fours, with elbows flexed, and putting its mouth against the infant's body on the ground. The open mouth may be pressed against any part of the infant's body, but especially its belly, back, head, neck, foot or arms." Similar to "Fore-limb crouch" (Fig. 8B of Hunt et al. 1996). Absent in bonobos? (Type A) Category 3–6. See video.

stare fixedly (STF)
Goodall (1989) Fixed stare: "The chimp stares intently at the individual with whom it is interacting...." At Mahale, groomee often stares fixedly at

groomer during intense social grooming. Also in threat, predation, and monitoring. Cf. <glance>, <look>, <peer>, and <watch>. Kano's (1998) Stare for bonobos, who stare mutually during mating. (Type A) Category 1. See video.

stare fixedly with head down and bottom up (SFH)

Infant female, Acadia, stood quadrupedal with head lowered onto palmer side of flexed lower right arm, resting on ground, while raising buttocks, and stared at mother being groomed by adult male. Request for suckling? Mother finally suckled Acadia. Weanling, Acadia, approached mother, hesitant to make contact with her, and showed this idiosyncratic posture. (Type A) Category 9. See video.

startle flee

See <flee after startle>.

startle flinch

See <parry>.

startle reaction

See <parry>.

steal (STL)

Remove surreptitiously object, without owner's awareness. See Goodall's (1989) Steal, which includes both <snatch> and <steal>. Also in bonobos (Kano unpublished). (Type C) Category 1.

stem pull-through

See <pull through stem>.

step on (STP)

Infant or juvenile elicits play by stepping on reclining partner. Often response of younger partner to older partner who lies supine in play-soliciting posture. Goodall's (1989) Ride onto. (Type A) Category 3–7. See video.

step over (STV)

Step over another lying on ground, who shows no response to stepping individual. No obvious function. (Type C) Category 3–7. See video.

step up on leg (STU)

Infant steps onto leg of mother in climbing up to ride on her back. (Type B) Category 2–5. See video.

stepping-stick (STE)

Branch used as "sandal" to avoid sharp thorns when moving in kapok tree (*Ceiba pentandra*). Reported from Tenkere (Alp 1997). Absent at Mahale. See also <seat-stick>. (Type B)

stingless bees

Small, colonial, honey producing bees of genus *Trigona*.
See <eat insect>.

stir water (STW)

Stir water vigorously with hand, index finger, large leaf, or foot in river. Solo play of juveniles, adolescents and adults. See also <play with water>. (Type B) Category 8. See videos. See also video "play with water."

stomp

See <stamp>.

store (STO)

Hold leaves in hand and start chewing only after collecting handful. Occurs when eating tough leaves of *Ficus exasperata*. Seen also at Gombe (Wrangham 1977). (Type A) Category 3–5. See video.

strike down prey

See <knock with both arms> or <knock with one arm>.

strip leaf (SPL)

Sit and tear off one or few leaves in quick succession from shrub or small tree. Older adolescent male, Orion, does this in courtship. Cf. <pull through in courtship> in which pull-through movement against stem is performed. Habitual threat gesture at Gombe and Kibale, and present at Bossou (Whiten et al. 1999). (Type A) Category 8. See video.

strip leaves by pull-through

See <pull through>.

stroke (STR)

Stroke back of another with extended fingers. Done by adolescent male, Cadmus. Idiosyncratic? Stroking face of another with fingers and palm is rare and of unknown function. Cf. <scratch socially, stroke type>. (Type A) Category 8–9. See video.

struggle free (SFR)

When pinned down by stronger partner in play, or by group during gang attack, victim exerts maximum effort to get free by <push>, <pull>, <kick>, <bite>, etc. (Type D) Category 1–3.

stumble (STM)

Lose footing, but rarely trip over stone or stick. (Type A) Category 1. See video. See also video "slap other."

subcutaneous tumor

Tumor-like skin swelling of hemispheric shape, approximately size of ping-pong ball, often protruding from lower abdomen of chimpanzee (Nishida, Fujita, Inaba et al. 2007). See video.

subgroup
See <party>.

subordinate
Lower-ranking of two individuals. Cf. <dominant>. See Fig. 11.

suck (SKK)
Suck nipple. Plooij's (1984) Suck (SVC). Goodall's (1989) Suckle (sic). Kano's (1998) Suckle (sic) for bonobos. (Type A) Category 1. See video.

suck in reassurance (SKR)
Goodall (1989) Reassurance suckle (sic): "A brief suck when the infant has been frightened or hurt." Absent in bonobos (Kano 1998). (Type C) Category 3. See video.

suck self (SKS)
Suck own nipples. Female who has lost newborn rarely sucks her nipples. Matsumoto-Oda's (1997) Self suckle. (Type A) Category 3–8. See video.

suck thumb (SKT)
Infant sucks own thumb, but rarely sucks other digits. (Type A) Category 1. See videos.

suck toe (SKO)
Infant sucks own toe like <suck thumb>. Adults and immatures suck toe to remove eggs of sand flea. See also <remove sand flea>. (Type A) Category 2–6. See video.

suckle (SUC)
Mother allows infant to suck from nipples. Goodall's (1989) Suckle. Kano's (1998) Suckle for bonobos. (Type A) Category 1. See video.

supplant (SPP)
Nishida and Turner (1996) Displacement. Displace another from food patch by <bark>, <approach>, <run>, <push>, <raise arm quickly>, or <shake branch>. Adult males usually supplant adult females and other adult males. Cf. <vacate>. In bonobos, adult females supplant adult males, who supplant other adult males (Kano 1998). (Type D) Category 2. See video.

support (SUP)
Intervene in fight and side with either of conflicting parties. Rare in bonobos (Kano unpublished). (Type D) Category 1. See video.

support dominant (SUD)
Intervene in fight and side with dominant party (Nishida and Hosaka 1996). de Waal's (1982) Winner support. Cf. <support subordinate>. (Type D) Category 1–3. See video.

support older (SUO)
Intervene in fight and side with older rather than younger party (Nishida and Hosaka 1996). (Type D) Category 1–6. See video.

support subordinate (SUS)
: Intervene in fight and side with subordinate party. de Waal's (1982) Loser support. Cf. <support dominant>. (Type D) Category 1–3. See video.

suspend (SPD)
: When chimpanzee suspended above ground, dangle by hand object such as carcass. (Type A) Category 1–3. See video.

suspend and shake up and down (SPS)
: When chimpanzee suspended above ground, dangle and shake object up and down by hand. Solo object play. (Type A) Category 3–8. See video.

swagger bipedal (SWB)
: Goodall (1989) Bipedal swagger: "In an upright or semi-upright posture the chimp sways, often rhythmically, from one foot to the other. The animal may remain in one spot to swagger or move forward during the swagger. The arms are normally held out from the body, the shoulders hunched up. It is often used during aggression, greeting, and courtship." van Hooff's (1973) Sway walk. Plooij's (1984) Bipedal swagger (BIS). Nishida's (1997) Bipedal swagger. Often performed as separating intervention just before charging display. Similar behavior occurs in bonobos of Wamba (Kano 1998). (Type A) Category 2. See video.

swagger on knuckles (SWK)
: Adult male, Pim, approaches estrous female in orthograde posture with knuckles on ground and penis erect. Idiosyncratic courtship display. (Type A) Category 8–9. See video.

swallow leaf (SWL)
: Wrangham and Nishida (1983) Leaf-swallow. Ingest slowly and without chewing, usually in early morning, leaves of certain species of herbs and trees. Leaves of *Aneilema* spp., *Aspilia mossambicensis*, *Commelina* spp., *Ficus exasperata*, *Lippia pulicata* and *Trema orientalis* swallowed at Mahale. Known from Taï (Boesch 1995), Kibale, Bossou, Gombe, and other localities (Huffman 1997). Species of swallowed leaves differ from place to place. Culture universal in chimpanzees? Reported for bonobos (Huffman 1997). (Type A) Category 2. See video.

S

swallow seed (SWS)
: Ingest seeds without chewing them. Includes tiny seeds such as fig (*Ficus* spp.) and large seeds (c. 3 cm long) such as *Pycnanthus angolensis*. (Type A) Category 1.

sway and move (SWM)
: Climb high up, sway tree trunk back and forth using body's weight, then catch and grasp terminal branch of another tree while original tree is bent over. Kano's (1998) Sway tree/branch for bonobos. Arboreal locomotion. Cf. <swing and take>. (Type A) Category 2. See video.

sway woody vegetation (SWV)

Component of charging display. While standing bipedal (sometimes, sitting), vigorously shake tree trunk of short tree/large horizontal branch/woody vine/ fallen trunk of tree/rock, back and forth with both hands, sometimes destroying the dead tree as result. Type of threat display and used also as courtship. Probably same as Goodall's (1989) Branch sway: "The chimp vigorously sways a growing branch usually when standing on one's feet. During a charging display, a male may pause and sway branches." Plooij's (1984) Branch-sway (BRS). Absent in bonobos of Wamba (Kano 1998). (Type A) Category 3–5. See videos.

swelling of sexual skin

Female anogenital region begins to swell cyclically at about 7 years of age and shows maximal swelling from about 10 years of age. After adolescence, sexual skin shows tumescence, maximal tumescence and dutemescence over typical menstrual cycle of 35 days. Maximal tumescence lasts 7–10 days, and in last two days of tumescence ovulation occurs (Graham 1981). Plooij's (1984) Swelling (SWE). See also <adolescent swelling> and <flabby bottom>.

swing (SWN)

Hang from woody vine by arms and swing back and forth. Social and locomotor play, solo play, threat, and charging display. If performer changes from one support to another, becomes <brachiate>. Goodall's (1989) Swing is equivalent to brachiation. Kano's (1998) Swing for bonobos. (Type A) Category 1. See video.

swing above (SWA)

Swing back and force above potential playmate while watching intently. Play solicitation, especially by infant female. (Type B) Category 8–9. See video.

swing and grasp (SWG)

Swing back and forth and when reaching farthest point, leap to or grab terminal branch of neighboring tree and transfer to it. Used when food abundant in canopy. Arboreal locomotion. Cf. <leap between trees> which lacks swinging. Also, cf. <sway and move>. (Type B) Category 3–6. See video.

S

swing and kick (SWC)

Hang from woody vine, swing, and forcefully kick buttress of tree, taking advantage of momentum of swinging movement. Sometimes, swing and kick other on back. Absent in bonobos of Wamba (Kano unpublished). (Type B) Category 3. See video. See also video "play solo."

swing forward and upward (SFU)

Hang by hands, extend legs forward by swinging, rotate whole body, and at same time assume orthograde position. Similar to human gymnastics. Arboreal solo play. (Type A) Category 1–9. See video.

T

tag (TAG)

Nishida (1983a). Alternately <chase> and <flee>. Includes <run>, <play-pant>, <play face> etc. Also in bonobos (Kano unpublished). (Type B) Category 1.

Taï

Study area of *Pan troglodytes verus* within Taï National Park, Cote d'Ivoire (5°52N′, 7°20′W). Research organized by C. Boesch from 1979 to present (Boesch and Boesch-Achermann 2000).

take (TAK)

Take food or tool with owner's knowledge and presumed consent. Modified from Goodall's (1989) Take. Tolerated scrounging equivalent to being <shared>. Cf. <snatch>, <steal> and <share food>. Plooij's (1984) Take away object (TAO). Also in bonobos (Kano 1992). (Type C) Category 1. See videos.

take dorsal

See <put dorsal>.

take finger in mouth (TFM)

Nishida (1983a), Goodall (1989): "Reach for the hand/foot of another and place part of hand/food or fingers/toes in its own mouth." Like <push finger into mouth>, reassurance behavior. <Take finger in mouth> much less frequent than <push finger into mouth> at Mahale. Absent in bonobos of Wamba (Kano 1998). (Type A) Category 3–5. See video.

take ventral

See <put ventral>.

tandem walk

See <follow in contact>.

tap heel (TAP)

Solicit copulation by hitting heel against ground/substrate. Courtship by males at Bossou (Nakamura and Nishida 2008). Shown by only one adolescent male, Orion, at Mahale. Idiosyncratic. (Type A) Category 9. See video.

teach (TCH)

At Mahale, teaching (as defined by Caro and Hauser 1992), if any, comprises only discouragement (Nishida 1987) or negative reinforcement. Mother or allomother only rarely removes and discards leaf from mouth of infant, if leaf is not in diet of group (Nishida 1983a; Goodall 1986). No systematic teaching (cf. Boesch 1991c). Teaching absent in bonobos of Wamba (Kano unpublished). (Type C) Category 8–9.

tear (TER)

Remove leaf from branch or stalk with hand. Common method of harvesting foliage. (Type A) Category 1. See video.

tease (TEA)

Dominant provokes subordinate by mild threats, such as throwing branch or "fishing", without recourse to violence. Different from Adang's (1984) Tease, which is similar to <harass>. At Mahale, adolescent female with monkey tail tried to "fish" for juvenile male: She carried colobus tail up tree and dangled it from above. Whenever male jumped for it, she withdrew it, so that he could not reach it. After several attempts, male got tail, and female grinned, suggesting that she was just trifling with male, not intending to give up tail. Adolescents and juveniles throw branches at other adolescents or juveniles, apparently to tease them. Cf. <trifle with> and <harass>. (Type D) Category 1–8. See video.

teeth-bare

See <grin>.

teeth-clack

See <clack teeth>.

temper tantrum

See <throw temper tantrum>.

Tenkere

Study area of *Pan troglodytes verus* in Outamba-Kilimi National Park, northern Sierra Leone (9°47′N, 12°0′ W). Research done by R. Alp (Alp 1997).

termite-fish

See <fish for termites>.

T

territory

Area occupied, patrolled, and defended by unit group or community. Cf. <core area>.

thigh support

See <transport with thigh support>.

threaten (THR)

Intention movement or anticipatory gesture of aggression. Goodall's (1989) Threat: "… a repertoire of gestures to elicit submissive behavior in the individual the gestures are directed toward… The repertoire includes <tip head>, <flail arm>, <hit toward>, <throw at>, <shake branch>, <stamp>, <slap>, <charge>, <display, charging>, <hunch>, <stare fixedly>, <bark>, <cough bark>, <waa bark>, <wraa>, <compress lips> and <scream> with <grin-full-open>". Kano (1998) lists for bonobos similar elements: "stare", "head-tip", "wrist-shake", "slap", "branch-sway", "arm-raise", "threatening approach", and "pseudo-charge". (Type B) Category 1–2. See video.

throw (THO)
Plooij (1984) THO. Goodall (1989): "Objects, such as stones, rocks, branches, sticks, handful of grass, etc. may be thrown underarm or overarm…" Bonobos also throw, but underarm only (Kano 1998). (Type B) Category 1. See video.

throw at (TAT)
Goodall (1989) Throw at: "When an object is aimed at a specific objective… aimed throwing". Throw object directionally. Customary at Gombe, Mahale, Taï, Bossou, and Goualougo (Sanz and Morgan 2007). Present at Assirik, Kibale and Budongo. Aimed throwing is divided into <throw at animate object> and <throw at inanimate object>. (Type B) Category 1–3.

throw at animate object (TAM)
Aimed throwing at animal, conspecific, prey such as bushpig, or competitor such as baboon and human. Rarely throw stone or branch at conspecific as intimidation or play and rarely hit other. Habituated chimpanzees of Mahale (Nishida et al. 2009) and Taï (Boesch 2009) throw branches at human for intimidation or play solicitation, but rarely hit. Throw at conspecific is customary at Goualougo (Sanz and Morgan 2007). Throw at prey not recorded at Mahale. Absent in Wamba bonobos (Kano unpublished), but Lomako bonobos threw fruit at tortoise (Hohmann and Fruth 2003). (Type B) Category 8–9. See video.

throw at inanimate object (TIN)
Aimed throwing at inanimate objects. Adult males at Mahale throw heavy stick or rock against wall of metal house to make loud noise, or into water (<throw splash> below). (Type B) Category 1–3. See video.

throw branch (THB)
Throw branch or stick in upright posture with arm. (1) Element of charging display by males and females, and (2) courtship pattern to estrous female by juvenile or older infant males (Nishida 1997). Older juvenile male once threw 1-m stick from distance of 8 m while pant-grunting to adult male. Bonobos throw branch forward at end of branch-dragging display (Kano 1998). (Type A) Category 2. See video.

throw dry leaves (THD)
In dry season, males of Mahale run, pick up pile of dry leaves from ground, and hurl them by hand. Occurs most frequently at climax or onset of

charging display, but also during pant-grunting. Juvenile male throws dry leaves as solo play while traveling. Not used in courtship. Absent in bonobos of Wamba (Kano unpublished). (Type A) Category 7. See video.

throw fruit against rock (THF)
At Gombe, throw hard-shelled fruit (e.g. *Strychnos*) against rock or tree trunk to crack it open (McGrew et al. 1999). Plooij's (1984) Banging (BAN). Absent at Mahale although *Strychnos* available. (Type A)

throw sand (THS)
Adult male, Fanana, throws sand with hand while running bipedally. Within M group's range, sandy ground is rare. Element of charging display and play. (Type A) Category 8–9. See video.

throw splash (THW)
Adult males of M group, Ntologi, Kalunde, Nsaba, Fanana, and Masudi, lift and throw heavy rocks into stream (cf. <throw at>) often with two hands, which produces loud splash that intimidates others (see Nishida 1993b). Special case of <throw stone or rock> or <throw at>. (Type A) Category 7. See videos.

throw stone or rock (TSR)
Hurl stones in charging display and also in play. Adult male, Alofu, simulated to hurl stone when his sex partner refused to follow him during herding. Bonobos of Wamba throw only branches because stones unavailable (Kano 1998). See <throw splash>. (Type A) Category 3–5. See video.

throw temper tantrum (TTT)
Goodall (1989): "The chimpanzee screams loudly, and may leap up, fling arms above his/her head and then slap them onto the ground or beat the ground with his hands. He may hold himself to the ground on his face, hug a tree, or himself. He may rush off, tumbling over and over, still screaming. The screaming often results in glottal cramps…" Plooij's (1984) Throw temper tantrum (TEM). Typically done by infants during weaning conflict, when mother rejects sucking, food sharing, transport, grooming etc. Adults sometimes show in response to rejection by dominants. At Mahale, infant had tantrum while hanging and swinging by feet, as if it were about to fall head first from high above ground. In temper tantrums, infant not only protests, but also appears to monitor how much it can induce care from rejecting mother. See <monitor mother>. Bonobos show similar behavior (Kano 1998). (Type B) Category 1. See videos.

thrust (THU)
Plooij (1984) THU. Goodall's (1989) Pelvic thrust: "Rhythmic back and forth movements of the pelvis… Occurs during copulation and… in reassurance mounting…" Two adult males embrace ventro-ventrally and bipedally in reassurance or reconciliation, and one thrusts against other. Adult females occasionally mount other females and thrust. Immatures thrust in play. May thrust on part of other's body such as leg or on object. Adolescent male, Cadmus, thrust against colobus carcass. Bonobo males

thrust like chimpanzees (Kano 1998). (Type A) Category 1. See video.

thrust bipedal (TBI)
Stand upright on feet and thrust in non-mating context. Courtship or reassurance shown by few adult and adolescent males. Intention movement of copulation likely as origin. (Type A) Category 8. See video.

thrust in vacuum (THV)
Thrust, but without any conspecific or object (*in vacuo*). Infants or juveniles do as solo play and during <play walk>. (Type A) Category 3–8. See video.

thrust, misdirected (THM)
Thrust not at rump, but on other body parts, such as leg, head, or shoulder, etc. Function is similar to usual thrust. (Type A) Category 3–7. See video.

thump
See <rap>.

tick
Ectoparasites of chimpanzees of genus *Ixodus*, etc.

tickle (TIC)
Plooij (1984) TIC. Goodall (1989): "The chimp puts one or both hands/feet on the body of the partner, usually between the neck and shoulder or in the groin, and makes tickling movements with the fingers." See also <poke> and <pat>. Bonobos may touch, push lightly, hold, or grasp infant or juvenile in social play, which may be equivalent to chimpanzee tickling (Kano 1998). (Type B) Category 1. See video.

tickle self with object (TSO)
Whiten et al. (1999) Self-tickle: Tickle self using objects. Habitual at Gombe. Present at Goulougo (Sanz and Morgan 2007). Absent at Mahale. (Type B)

tickle with hand or foot
See <tickle>.

tickle with mouth
See <mouth>.

T

tilt head (TIH)
Incline head to one side while standing quadruped, to solicit play, or to look above. (Type A) Category 3–8. See video.

tip head (TPH)
Goodall (1989) Head tip: "… a threatening gesture. Head is jerked very slightly backwards, at the same time chin is raised. The performer faces the individual being threatened and the gesture is usually accompanied by a cough bark and often an arm raise." Plooij's (1984) Tip (TIP). Also in bonobos (Kano 1998). (Type A) Category 2.

Tongo

Study site of *Pan troglodytes schwein-furthii* within Virunga National Park, Democratic Republic of Congo (1°20′–1°15′S, 20°05′–29°10′E). Research done from 1987 to 1990 (Lanjouw 2002).

tool

Object extracted or detached from environment and used to obtain benefits (e.g., raise efficiency of performance) or pleasure, in aid or extension of manipulable organs such as limbs and mouth (see also Beck 1980). See <make tool> and <use tool>. See Fig. 36.

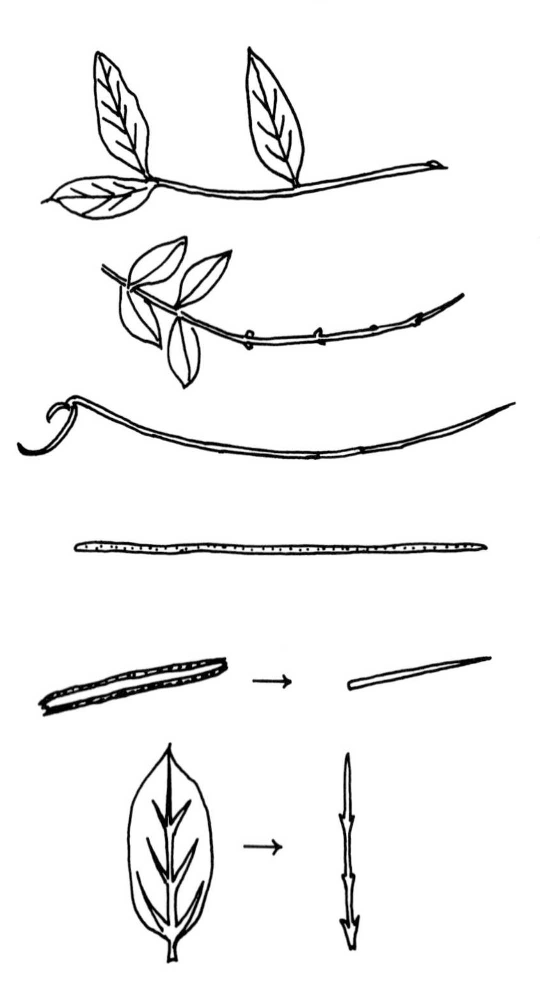

Fig. 36 Tool: Tools used by K group to fish for *Camponotus* ants (T. Nishida)

tool composite (TCM)

Two or more tool types used simultaneously and complementarily, to achieve single aim, e.g. hammer and anvil. Human examples are bow and arrow, mortar and pestle, etc. Absent at Mahale. (Type D)

T

tool set (TST)

Two or more tool types used in obligate (not flexible) sequence to achieve single aim (Brewer and McGrew 1990). Sugiyama's (1997) Tool composite. Includes <push and pull sponge> (Sugiyama 1995b), <pound pestle> and <leaf-sponge> (Sugiyama 1995b), <hammer nut> and <pick out nutmeat> (Boesch and Boesch 1990), <puncture> and <fish for termites> (Fay and Carroll 1994; Suzuki et al. 1995; Bermejo and Illera 1999; Deblauwe et al. 2006; Sanz and Morgan 2007), <pound>, <lever open> and <dip fluid> (Sanz and Morgan 2007). At Mahale, two cases of using tool-set seen when immatures retrieved leaf-sponge from tree-hole with stick (Matsusaka et al. 2006). Cf. <tool composite>. (Type D) Category 1–3. See videos (Goualougo).

tool use

See <use tool>.

touch (TOU)

van Hooff (1973) Touch. Plooji (1984) TOU. Nishida (1983a) Touch. Goodall (1989): "A chimpanzee reaches out with a hand (or occasionally foot) and touches another, with fingers (or toes), by laying the whole palmar surface of the hand on the other's body…" Touch signals reassurance to subordinate from dominant, and appeasement to dominant from subordinate. Mother touches infant to signal departure. At Mahale, estrous females approach and touch males on shoulder, etc., to solicit copulation. To touch head of another sometimes signals to groomee by groomer to change posture. Parts frequently touched are mouth, chin, head, hand, shoulder, and genitals. Mouth most often touched by adult male, Kalunde, to appease alpha male, Fanana. See also <push finger into mouth>, <groom>, <inspect>, <pat> and <push>. Adult bonobos much less often touch each other, mostly done by infants and juveniles (Kano 1998). (Type A) Category 1. See video.

touch fruit (TOF)

Touching hard-shelled fruit important source of information about ripeness. Sometimes, only touching suffices, but may be sniffed and bitten. Type of <inspect fruit>. Absent in bonobos of Wamba. (Type A) Category 3–6. See video.

touch scrotum (TOS)

In mounting, mounted male often reaches back to touch genital area of mounter with hand, while screaming loudly. Adult male also touches scrotum of another for reassurance. While pant-grunting to adult male, adult female often reaches to touch scrotum. Kortlandt's (1964) Caress another's scrotum, for chimpanzees of Beni. (Type A) Category 1–5. See video. See also video "sniff finger."

touch with foot (TWF)

Function is similar to <touch with hand>, but rare. Juvenile "extends its leg" and clandestinely touches any body-part of newborn (often sibling) while grooming mother's back. (Type A) Category 1–3. See video.

touch with hand

See <touch>.

toy

Object used for solo or social play. Object extracted or detached from environment used for play solicitation or as focal object during social play becomes tool. Comprises branch, fruit, leaf, flower, stone, sand, animal, piece of animal skin, piece of clothes, earthenware, cardboard, etc.
See video.

toy with

See <trifle with>.

tradition

Behavior learned socially (versus individually), shared by at least most members of at least one age or sex class, and transmitted from generation to generation. See also <culture> and <fashion>.

trample (TRL)

Plooij (1984) Trample (TRA). Put foot or feet on body of reclining other. Sometimes accompanied by stamping. Rare pattern of social play.
Cf. <sit on> and <step on>. (Type B) Category 3–8. See video.

trampoline

See <hang and stamp>.

transfer (TRS)

Migrate permanently from one unit group to another (Nishida and Kawanaka 1972). Females normally transfer from natal group to another during adolescence- at age 11 at Mahale (Nishida et al. 1990) but 12–13 at Taï (Boesch 2009). No sexually mature males transfer at Mahale, thus, chimpanzees are "patrilocal" (Nishida 1979). Also for Gombe (Goodall 1986), Taï (Boesch and Boesch-Achermann 2000), Ngogo (Mitani et al. 2002) and the bonobos of Wamba (Kano 1992), although female emigration varies across sites from 50 to 90%. Rarely, mother with juvenile offspring transfers (Nishida et al 1985; Takahata and Takahata 1989). Adult male bonobos rarely tranfer (Hohmann 2001). See <emigrate> and <immigrate>. Cf. <visit>. (Type D) Category 1. See video.

T

transport (TRP)

Nishida (1983a): Quadrupedally carry infant on belly or on back. Carry food between lips, in hand or foot, on back or head, in groin pocket or neck pocket, or drag food by hand. During play, juveniles often transport objects such as twig, fruit, ant-nest etc. McGrew's (1992) Play-start.
See <solicit play with object in mouth> and <ride>. Bonobos also transport food similarly (Kano 1998). (Type B) Category 1–2. See video.

transport bipedal (TRB)

When carrying much food or infant, walk on feet only. Bonobos carry sugar cane bipedally (Kano 1998). (Type A) Category 1. See video.

transport corpse of infant (TRC)

Mother carries dead body of infant by hand, neck or groin pocket, or on back. May persist for 3 months, and corpse becomes mummy. Known from Bossou (Matsuzawa 1997), Ndoki (Kuroda 1998), Gombe, and Mahale. Also in bonobos (Kano 1992). (Type B) Category 2. See video. See also video "descend."

transport food (TFO)

Transport of food occurs when: (1) individual (especially youngster) seeks to follow companion who has eaten and departed, (2) (especially heavy adult male) seeks to eat in safe position in tree, and (3) seeks to eat in shade. Bonobos also carry food (Kano 1998). (Type B) Category 1. See video. See also video "transport in hand."

transport in foot (TRF)

Carry grasped in foot object such as fruit, branch, stone, carcass, etc., usually in tree. Rare. (Type A) Category 2–3. See video.

transport in groin pocket (TGP)

Transport object clamped between thigh and abdomen. Goodall's (1989) Groin pocket. See also <hold object in groin pocket>. Kano's (1998) Groin pocket for bonobos. (Type A) Category 2. See video.

transport in hand (TRH)

Common means of transport by hand of object such as fruit, branch, stone, carcass, etc. Stone carrying occurs during solo and social play, and occasionally before intimidation display or attack by stone throwing. (Type A) Category 1. See video.

transport in mouth (TRM)

Carry object by mouth (between lips or teeth). Hold in mouth vines or twigs as materials of fishing tools and transport to fishing sites. Hold in mouth food such as leafy twigs, fruit during transport, or carcass. Juvenile transports leafy branch in mouth to solicit play (see <solicit play with object in mouth>) and during play (<toy>). Mother, Xtina, held infant's nape or back in mouth and transported for several months. Adolescent male, Darwin, also did so few times (Nishida et al. 2009). Recorded at Bossou (Sugiyama 2008). Bonobos carry objects in mouth (Kano 1998). (Type A) Category 1. See videos.

transport in neck pocket (TNP)

Goodall (1989) Neck pocket, Plooij (1984) ONP. Transport object clamped between lower jaw and shoulder or chest. See also <hold object in neck pocket>. Rare in bonobos (Kano 1998). (Type A) Category 2. See video.

transport on back (TRD)

Carry dorsally object such as branch, carcass, piece of skin, etc. Infant carried on back from 3–6 months after birth. (Type A) Category 2–3. See video.

transport on head or nape (TOH)

Plooij (1984) OOH. Carry piece of skin or carcass balanced or draped on head. Mother, Chausiku, regularly carried newborn on head. Juvenile male, Michio, carried stone on nape in solo play. See also <hold object on head>. (Type A) Category 8. See video. See also video "transport corpse of infant."

transport on shoulder (TSH)

Carry large object such as carcass on shoulder. (Type A) Category 1–3. See video.

transport quadrupedal (TRQ)

Transport object with hand, in mouth, neck pocket, on shoulder, or on head, while walking on four limbs. (Type B) Category 2.

transport two offspring (TRO)

Mother simultaneously carries juvenile offspring on back and infant offspring on belly, or rarely carries both on back. Occurs in dangerous situations such as crossing large river. Bonobo mothers seem to carry two offspring more often than do chimpanzee mothers, as birth interval is shorter (Kano 1998). (Type B) Category 1–2. See video.

transport with hand support (TWH)

Goodall (1989) Hand support. Plooij's (1984) Support (SUP). Mother supports infant's back with hand. Newborn or infant weekened by injury or illness typically carried this way, as gripping power of lower limbs is weak. Sometimes seen during <ride dangling>. Kano's (1998) Hand support for bonobos. (Type A) Category 1–2. See video.

transport with thigh support (TRT)

Goodall (1989) Thigh support: "Method of supporting infant especially newborn, where the mother walks or runs with her thighs flexed, thus providing support for the infant's back. Usually the mother moves with rounded back and takes short steps." Mother also carries newborn in flexed lower limbs when brachiating or hanging in tree. Kano's (1998) Thigh support in bonobos. (Type A) Category 2.

travel (TRV)

Go from one place to another. Goodall's (1989) Travel. Plooij's (1984) Travel (TRV). Include <walk quadrupedal>, <walk bipedal>, <walk tripedal>, <wade>, <crutch>, <run>, <leap>, <somersault>, <climb>, <brachiate> and <descend>. Kano's (1998) Travel in bonobos. See Susman et al. (1980) for bonobo locomotor patterns. (Type D) Category 1. See videos.

travel alone after childbirth (TRA)

Days or weeks after parturition, mother with newborn spends time alone, apart from other group members, probably avoiding harassment. At Wamba, bonobo mother often accompanied by one or two males directly after giving birth (Kano 1998). (Type D) Category 2–3.

travel and play (TVP)
Travel while playing socially or solo in locomotor-rotational play such as somersault and pirouette. During travel youngsters occasionally find something interesting such as hole in the ground, slope with full of dry leaves, animal carcass, etc. which elicits play. (Type D) Category 2–3. See video.

trifle with (TRW)
Tease for amusement captured small animal such as hyrax (Hirata et al. 2001), squirrel (Zamma 2002b), or leopard cub (Hiraiwa-Hasegawa et al. 1986) by <hit>, <slap>, <flail>, <stamp>, <drag>, <drop>, etc. Hirata et al.'s (2001) Toy with. Victim sometimes killed in process. <Play face> shown suggests that trifling sometimes fun (Zamma 2002b). Cf. <tease> and <abuse carcass>. (Type D) Category 1–3. See video.

try play
See <solicit play>.

tug of war
See <pull object from opposing sides>.

tumble (TUM)
Lose balance and fall down slope. (Type B) Category 1–3. See video.

turn around (TUR)
Turn back to approaching individual (for example, seeking to beg). Goodall's (1989) Turn away: "Mother may turn from her infant to prevent suckling or begging. Adults may also turn their backs on beggars." Kano's (1998) Turn back in bonobos. (Type A) Category 1. See video.

turn away
See <turn around>.

turn face away (TFA)
Turn face away from partner who approaches to give kiss. (Type A) Category 1–8. See video.

turn face downward
See <lower head>.

turn face upward (TFU)
Incline head backward (extend neck), while being groomed on face. Also in bonobos. (Type A) Category 1–2.

turn round
See <look back>.

turn up lip (TUL)

Evert upper lip over muzzle, exposing upper gums and teeth for grooming. Cf. <flip lip>. (Type A) Category 3–8. See video.

twist (TWI)

When flesh difficult to detach by pulling from carcass possessed by another, wrench bone so that piece of carcass is detached. Mentioned in Boesch (2009), but not in Goodall (1989). (Type A) Category 1–3. See video.

U

Ugalla

Study area of *Pan troglodytes schweinfurthii* in Tongwe East Forest Reserve, Tanzania (5°19′S, 30°37′E). Research organized by J. Moore from 1985 to present (Moore 1996). Also, Research conducted by T. Kano, H. Ogawa and G. Idani from 1991 to present. Another current research by A. Hernandez, A. Piel, and F. Stewart. Pioneer studies done by J. Itani and T. Kano in 1966–1967 and T. Nishida in 1975.

unit group

Social unit of chimpanzees (Nishida 1968), also called "community" (Goodall 1973). Multi-male, multi-female group with group size of up to 150, occupying territory 10–200 km^2 in area (Mitani et al. 2002). Adult females outnumber adult males. Shows fission and fusion grouping pattern. Females emigrate from natal unit group at age of about 11 years. Inter-group relationships are antagonistic. Unit group also main reproductive unit and extra-group paternity is rare at Gombe (Constable et al. 2001), Taï (Vigilant et al. 2001) and Mahale (Inoue et al. 2008). Bonobos have unit group of similar size, composition and male philopatry (Kuroda 1979; Kano 1982a; Furuichi 1989), but inter-group relations are peaceful (Idani 1990) and even fusion of two groups is indicated (Hashimoto et al. 2008). See Fig. 37.

uproot

See <pull out>.

upside-down suspension

See <hang upside-down by feet> and <hang upside-down by hands>.

urinate (URN)

Excrete urine. (Type B) Category 1. See video.

urinate, prone (URP)

Lies on belly and excretes urine. (Type A) Category 2–8. See video. See also video "defecate, prone".

urinate quadrupedal (URQ)

Female excretes urine in standing quadrupedal posture. Male does with adducted thighs, sometimes while walking. (Type A) Category 2–3. See video.

Fig. 37 Unit group: Procession of M group, more than 60 chimpanzees at one time (T. Nishida)

urinate, sitting (USP)

Urinate in sitting posture. (Type A) Category 1–3. See video.

use tool (UST)

Use detached object to change state or position of another (goal) object. At Mahale includes <fish for carpenter ant>, <fish for termite>, <clip leaf>, <probe>, <drag branch>, <club>, <flail long object>, <throw stone or rock>, <solicit play with object in mouth>, <toy>, <probe nasal passage>, <squash leaf> and <make bed>. Great local variation exists in tool-use (McGrew 1992; Sugiyama 1997; Yamakoshi 1998; Watts 2008b). Tool-using repertoire of bonobos is limited: <cover from rain>, <drag branch>, <throw branch> and <make bed> at Wamba (Ingmanson 1996; Kano 1998). (Type D) Category 1–3. See video. See also video "expel".

U

V

vacate (VAC)

Upon approach of more dominant individual, one of grooming party leaves partner and vacates position to dominant individual. Subordinate individual also may leave feeding patch when dominant approaches.
Cf. <supplant>. (Type D) Category 1–3. See video.

ventral riding

See <ride ventral>.

ventro-dorsal copulation

See <copulate dorso-ventral>.

ventro-ventral copulation

See <copulate ventro-ventral>.

ventro-ventral embrace

See <embrace full>.

vertical climb

See <climb vertical>.

vertical leap

See <leap down>.

vertical slide

See <slide down vertically>.

vigilance (VIG)

Visual monitoring to collect information on possible threat of predation or of social conflict among conspecifics (Kutsukake 2006, 2007). See <glance>, <look>, <monitor>, and <watch>. (Type D) Category 1.

visit (VST)

Migrate temporarily from one unit group to another. Adolescent and young nulliparous adult females in estrus visit neighboring groups, then return to natal group when sexual skin deflates. Known from Gombe (Goodall 1986),

Taï (Boesch and Boesch-Achermann 2000) and Mahale (Nishida et al. 1990). Differs from <transfer>, which is permanent. (Type D) Category 1–2.

vocalize (VOC)

Call. Many calls are specific to contexts and express emotions. Includes <bark>, <choke in tantrum>, <cough bark>, <grunt>, <hoo>, <hoot>, <huu>, <pant>, <pant-bark>, <pant-grunt>, <play-pant>, <scream>, <squeal in copulation>, <staccato call>, <waa bark>, <whimper>, and <wraa>. Behavior accompanying vocalizations includes: <clack teeth>, <cough>, <fart>, <hiccup>, <sputter>, <smack lip>, <sneeze>, etc. Cf. <make sound>. (Type D) Category 1.

vomit (VOM)

Plooij (1984) KOT. Rare, but infants do more often. See also <reingest vomit> and <sickness>. (Type A) Category 1. See video. See also video “reingest vomit.”

waa bark (WAA)

Goodall (1989): “Loud, sharp sound given in a variety of agonistic contexts. It is usually accompanied by an arm threat or more vigorous gestures.” See also <bark>. (Type A) Category 3. See video.

wad

See <wadge>.

wade (WDE)

Minimize contact with water, e.g. in river crossing, and use stepping stones if available, when trees on both sides of river too far apart for overhead travel. If stepping stone unavailable, walk quadrupedal or bipedal in river, according to water depth. Cf. <leap bipedal> and <leap quadrupedal>. Bonobos wade similarly (Thompson 2002). (Type A) Category 1–7. See video.

wadge (WDG)

Chew and compress food between lower lip and teeth or between tongue and palate. Includes <wadge without adding leaf> and <wadge by adding leaf>. Neither Goodall (1989) nor Nishida et al. (1999) differentiated two forms. See also <squeeze>. Bonobos also wadge (Kano 1998). (Type B) Category 2–3. See videos.

wadge by adding leaf (WDA)

Chew and compress meat, skin, honey, grub, egg, semen, or fruit with leaves between lower lip and teeth, or between tongue and palate. After juice extracted, remaining inedible parts such as hair, eggshell, seed, leaf, etc. spat out. Chewed leaves from species not eaten solely, e.g. *Saba comorensis*, and even dry leaves are used. Fruits frequently wadged in this way at Mahale include *Canthium crassum*, *Harungana madagascariensis*, *Parinari curatellifolia*, *Pseudospondias microcarpa*, *Psychotria peduncularis*,

Syzigium guineense, *Uapaca kirkiana* and *Uapaca nitida*. Leaves chewed with fruit usually come from trees of eaten fruit. (Type A) Category 3. See video.

wadge without adding leaf (WDP)

Extract juice by chewing and compressing some fruit, stalks of grasses or herb vines or young leaves between lower lip and teeth, or between tongue and palate. Unlike <wadge by adding leaf>, leaves (except main edible leaves mentioned above) are not added. After juice extraction, remaining inedible parts spat out. Piths of grasses such as *Pennisetum purpureum*, herbaceous vines such as *Ipomoea rubens*, or young leaves of *Brachystegia spiciformis* chewed and sucked with wadge repeatedly taken out and returned to mouth. (Type A) Category 2. See video.

wait for companion (WIT)

Goodall (1989): "An individual sets off, then looks back at a companion. If the latter is not following, the first stops to wait…" Ntologi, alpha male, often waited for adult males to follow. Bonobos wait similarly (Kano 1998). (Type D) Category 1. See video.

wait turn (WTR)

Wait for predecessor to vacate position, when target food (such as carpenter ants) obtainable only from specific site (versus supplanting or displacing predecessor). Wait to groom target individual until another departs. Includes <sit> and <stand quadrupedal>. (Type D) Category 1–3. See video.

waiting posture

See <sit> and <stand quadrupedal, heel up> for <wait for companion>, <sit> and <stand quadrupedal> for <wait one's turn>.

walk backwards

See <retreat>.

walk bipedal (WAB)

Goodall (1989) Bipedal walk. Plooij (1984) WAB. Walk upright with much food in arms, on muddy terrain, or to initiate charging display. Susman et al.'s (1980) Bipedal walk in bonobos. (Type A) Category 1. See video. See also video "stand bipedal".

walk in sloth position (WSP)

Hang underneath branch using all hands and feet or combination of any three and locomote forwards. At Mahale part of solo play by immature chimpanzees. Bonobos also travel in this posture, which Kano (1998) called “Sloth-walk.” (Type A) Category 2. See video.

walk lame
See <limp>.

walk quadrupedal on backs of hands (WQB)
Rare solo locomotor play. (Type A) Category 8–9. See video.

walk quadrupedal on knuckles (WQK)
Goodall (1989) Knuckle walking, Plooij (1984) WAQ. Quadrupedal walk with knuckles on ground or large horizontal bough. Kano’s (1998) Walk. (Type A) Category 2. See video.

walk quadrupedal on palms (WQP)
Walk quadrupedal, with open palm on thin horizontal branch. Rare when walking on ground. Susman et al.’s (1980) Palmigrade quadrupedalism. (Type A) Category 2. See videos.

walk stealthily
See <creep>.

walk tripedal (WQT)
Walk with object in hand or walk with hand on back of another. Cf. <embrace half>. (Type A) Category 2. See video.

Wamba
Study area of *Pan paniscus* in the Equatorial Province of DRC (0°11′08″N, 22°37′58″E), also called “Luo Scientific Reserve”. Research organized by T. Kano from 1973 to 1991 and from 1994 to 1996 (Kano 1992). After civil war, research reorganized by T. Furuichi from 2002 to present (Furuichi and Thompson 2008).

wash (WAS)
Seen only as <wash colobus skin> below. Wash includes whole process of <dunk>, <rinse>, and <stamp>. At least six bonobos of Lilungu wash aquatic herb or earthworms in water before consuming them (Bermejo et al. 1994). (Type B) Category 9.

wash colobus skin (WAF)
Old adult male, Musa, once repeatedly dunked colobus pelt into running water, kept it immersed, rinsed it, stamped on it on rock, repeating process several times. Musa then nibbled skin (Nishida 1993b). (Type B) Category 8–9. See Fig. 38.

Fig. 38 Wash colobus skin: Adult male, Musa, washed colobus skin in stream. Four chimpanzees watched rare activity (T. Nishida)

watch (WAT)

Goodall (1989): "Gaze intently at what another is doing. If the face of the watching individual is very close to the other this is described as peering" (cf. <peer>). van Hooff's (1973) Watch. Cf. <stare fixedly>, <glance>, <look>. Kano's (1998) Watch in bonobos. (Type A) Category 1. See video.

watch water surface

See <look at water>.

wean (WEA)

Goodall (1989) Wean. Patterns of maternal rejection that encourage cessation of sucking and independence of infant from mother: Include <drop infant>, <go ahead>, <ignore>, <reject infant>, etc. Kano's (1998) Wean in bonobos. (Type D) Category 1. See video.

wedge (WEG)

Infant forces itself between two others interacting socially, often mother and grooming or mating partner. See also <interfere copulation>. (Type C) Category 3–5. See video.

whimper (WHP)

Goodall (1989): "A whole series of soft, low pitched sounds, rising and falling in pitch, which may lead to crying and screaming…" Whimper by male infant may elicit presentation by estrous female, so serving as courtship display (Nishida 1997). Estrous female may whimper to adult male for copulatory solicitation. Cf. <hoo>. Bonobos call similarly (Kano 1998). (Type A) Category 2. See video.

whimper-scream (WHS)

Goodall (1989) Cry: “A combination of loud whimpers and tantrum screaming. This is commonly heard when an infant is separated from mother”. At Mahale, weaning infant, especially if weaned when unusually young (e.g. 3 years old), whimper-screams continuously throughout mother’s estrous periods. Lost call of bonobos seems distinct from that of chimpanzees in acoustic nature, and both mature and immature individuals emit it (Kano 1998). (Type A) Category 2–3. See video. See also video “throw temper tantrum”.

whisk fly with arm (WHK)

Shake arm to keep flies from converging on self or object, especially when mother carries dead infant. Cf. <catch with hand>. (Type B) Category 3. See video.

whisk fly with leafy twig (WHL)

Sugiyama (1968) Fly whisk: Use leafy twig to fan away flies. Habitual at Taï and Budongo, present at Gombe (Whiten et al. 1999). Recorded for Lomako bonobos (Hohmann and Fruth 2003). Absent at Mahale. (Type A)

winner support

See <support dominant>.

wipe with detached object (WPD)

Whiten et al. (1999) Leaf-napkin: Use leaves to clean body surface. Goodall (1989): “Use of leaves, straw, paper, etc. to wipe dirt (feces, urine, mud, ejaculate, etc.) from self or others…” Very rare at Mahale, with only handful of cases of rump-wiping with leaf. Customary at Kibale. Present at Assirik, Taï, and Goualougo. Penis-wiping by male after copulation customary at Gombe (Goodall 1989) and Budongo (O’Hara and Lee 2006), but absent at Mahale. Wiping dirt from others not seen at Mahale. Cf. <rub muzzle>. Seen three times in bonobos at Wamba (Ingmanson 1996). (Type A) Category 8. See video.

wipe with still-attached object

See <rub muzzle>.

wraa (WRA)

Goodall (1989): “---a long drawn out, pure sounding call, neither high or low pitched… Given by chimpanzees when confronted with new or disturbing objects in the environment - a human before they were habituated to humans, sometimes buffalo, python, dead bush pigs, etc…” See also <bark>. (Type A) Category 3–5. See video.

wrestle (WRE)

Goodall (1989): “Two or more youngsters grab hold of each other and often roll over as they bite, tickle, kick etc. without losing contact.”
Cf. <play, rough and tumble>. Kano’s (1998) Wrestling play in bonobos. (Type B) Category 1. See video.

wrestle bipedal (WRB)

Wrestle in bipedal stance like Japanese "sumo". Lasts less than 5 s before return to usual <wrestle> in sitting and reclining postures. (Type B) Category 3–7. See video.

wrestle with fingers (WFI)

van Hooff (1973) Handwrestle. Goodall (1989): "Gentle fondling, holding, squeezing, tickling of one hand (occasionally foot) of intended play partner. This is how a mature individual typically seeks to initiate play with another adult. It may lead to a vigorous bout of social play." At Mahale, initiates social play in adult males and between adult males and females; overall most common pattern of social play among adults. Absent in bonobos (Kano 1998). (Type A) Category 3. See video.

wriggle (WRG)

Writhe or squirm from being heavily bitten by ants, etc. (Type B) Category 1–7. See video.

wrist towards

See <reach wrist toward>.

Y

Yalosidi

Study area of *Pan paniscus* in Equateur Province of DRC (1°50′S, 23°15′E). Research done by T. Kano and colleagues from 1973 to 1977 (Kano 1983; Uehara 1990).

yawn (YAW)

Goodall (1989) Yawning. Plooij (1984) YAW. Kano (1998) Yawn for the bonobos. (Type A) Category 1. See videos.

Discussion

Our list contains 891 behavioral patterns of Mahale chimpanzees, 45 patterns that have been recorded at other sites but not at Mahale, 91 nouns concerning chimpanzees and their study sites, and 357 synonyms, totaling 1,384 terms. Thus, we have added more than 500 new terms to the previous publication (Nishida et al. 1999). The last publication was the product of research that spanned more than 30 years. Why so many patterns have been added must be explained.

First, we did not use videographic records extensively for the last publication. Although we referred to video records taken by ANC Production and East Company, they were not major sources for the behavioral patterns. This time videotapes that add up to more than 1,500 h since 1999 were played back and watched again and again. This procedure has brought a qualitatively, as well as quantitatively, higher level of data scrutiny than before.

Second, the multi-authored article (Whiten et al. 1999) that was published simultaneously with our last paper in 1999 provided fruitful discussion with many colleagues from different research sites directly or via email and alerted us to subtle differences in behavioral patterns.

Third, and most important, the behavioral flexibility of chimpanzees is responsible for major revision of this paper. A comprehensive dictionary of human actions would amount to several thousand pages because human behavior is flexible and variable. If we take the number of chimpanzees we have intimately observed – only 250 individuals or so – for consideration, 180 pages for this encyclopedia would be remarkable. As we pointed out already, behavior as simple as social scratch is not only absent at some sites, but also the behavioral pattern of social scratch differs across sites where it exists. Chimpanzees are "cultured" (McGrew 2004) animals whose behavior is modified during maturation after birth, but also changes even during adulthood. This was illustrated by the changes shown by migrants in the features of the grooming hand-clasp (McGrew et al. 2001; Nakamura and Uehara 2004). From the viewpoint of cultural diversity, we conclude that we have currently reached only the starting point of understanding of behavioral diversity and flexibility of chimpanzees.

Currently it is difficult to classify behavioral patterns into Categories 1–9 since detailed glossaries of behavior are available from only two populations of chimpanzees, Gombe and Mahale, both belonging to subspecies of *P. t. schweinfurthii* and

T. Nishida et al., *Chimpanzee Behavior in the Wild: An Audio-Visual Encyclopedia*, DOI 10.1007/978-4-431-53895-0_10, © Springer 2010

one population of bonobos at Wamba. In this paper, we tentatively allocated categories to all the behavioral patterns on the basis of the "minimal status" and "maximum status". For example, <fish> is known from all subspecies of chimpanzees, so it is very likely to be a "chimpanzee universal" (Category 3), but it also is known from some indigenous humans, while it still unknown from bonobos. In this case, Category 3 is the minimal status of <fish>, and Category 1 (characteristic of LCA) is a possible maximum status. Then, we allocate <fish> to Categories 1–3 (Table 1).

Since the minimum status is the only category that has so far been confirmed for each behavioral pattern, we here use it as the principle means of summarizing our

Table 1 Numbers of behavioral patterns in each category and its phylogenetic position

Categories	No. of patterns	Subtotal	Phylogenetic positions
1	264	264	Likely shared by LCA
1–2	30	157	Likely shared by common ancestors of Pan
2	127		
1–3	50	125	Likely shared by *Pan troglodytes*
2–3	31		
3	44		
1–5	12	79	Likely shared by *Pan troglodytes schweinfurthii*
2–5	5		
3–5	58		
4–5	2		
5	2		
1–6	9	43	Likely to be Mahale customs
2–6	5		
3–6	24		
6	5		
1–7	17	77	Likely to be M group customs
2–7	3		
3–7	34		
5–7	1		
7	22		
1–8	13	92	Likely to be habits of a few M group chimpanzees
2–8	4		
3–8	26		
7–8	5		
8	44		
1–9	2	54	Likely to be idiosyncracies of M group chimpanzees
3–9	2		
8–9	31		
9	19		
Total	891	891	

results. Thus, we have obtained 263 behavior patterns likely shared by the common ancestors of *Homo* and *Pan*, 157 patterns likely to be shared by commons ancestors of chimpanzees and bonobos, 126 patterns likely to be chimpanzee universals, 79 patterns likely shared by eastern subspecies, 43 patterns likely to be Mahale cultures, 77 patterns likely to be M group cultures, 93 patterns likely to be habitual among a few M group chimpanzees, and 54 patterns likely to be idiosyncracies of M group chimpanzees (Table 1).

We are sure that the last common ancestors of *Homo* and *Pan* had such characteristics as drag branch, play aeroplane, female transfer, and many other patterns, but we are less sure that they engaged in patrol, ostracize, fish and many other patterns.

Probably, more important, we found many candidates of cultural behavioral patterns, namely 120 patterns (43+77), in addition to those 44 patterns that have not been recorded at Mahale but have been seen elsewhere. Although this needs confirmation by future detailed, long-term studies, we speculate that like current apes and humans, our common ancestors enjoyed rich cultural diversity in addition to common basic genetic tendencies.

However, most impressively, chimpanzees show a tremendous extent of diversity of individual or idiosyncratic behavioral patterns: 147! (93+54). This is the ethnographic diversity of chimpanzees that deserves much more attention as a source of new patterns (Nishida et al. 2009).

We hope that by viewing the video clips carefully as well as reading the glossary, researchers at other sites will discover much new material for discussing the behavioral diversity of chimpanzees living in the natural environment. Rich flesh will be put on the bones of Table 1 and full interpretation will be given only after detailed information is gathered in many other sites, in particular, for *Pan troglodytes troglodytes* and *P. t. vellerosus* and *Pan paniscus*.

Acknowledgments We thank Tanzania Commission for Science and Technology, Tanzania Wildlife Research Institute, and Tanzania National Parks for permission to do the field research, and Mahale Mountains National Park and Mahale Mountains Wildlife Research Centre for logistic support. Our gratitude goes to Hosea Y. Kayumbo, Costa Mlay, George Sabuni, Charles Mlingwa, Erasmus Tarimo, Edeus Massawe, A.K. Seki, M. Mbaga and their staff for their friendship and encouragement; We thank Michio Nakamura, Gaku Ohashi, Tetsuya Sakamai, Crickette Sanz and Dave Morgan for allowing us to use their precious video clips; Tamotsu Asou, Richard Byrne, Satoshi Hirata, Michio Nakamura, Yukimaru Sugiyama, Rikako Tonooka, Caroline Tutin and Moyo Uehara for allowing us to use their precious photos or illustrations; Tamotsu Asou, Miho Nakamura, Mitsue Matsuya and the late Masayasu Mori for allowing us to use their videos for study; Toshimichi Nemoto and Asami Kanayama for helping us in various ways in Tanzania. We thank our colleagues who have studied chimpanzees at Mahale for sharing information on the behavioral patterns, in particular, the late Kenji Kawanaka, the late Shigeo Uehara, Hitoshige Hayaki, John Mitani, Linda Marchant, Kazuhiko Hosaka, Akiko Matsumoto-Oda, Michio Nakamura, Noriko Itoh, Tetsuya Sakamaki and Nadia Corp. We are indebted to our field assistants, in particular, Ramadhani Nyundo, Rashidi Kitopeni, Rashidi Hawazi, Mtunda Hawazi, Mosi B. Kasagula, Hamisi B. Kasagula, Mosi Hamisi, Kabumbe Athumani, and Mosi Matumla. We are indebted to Naomi Miyamoto for assistance in editing video clips. We thank Frans de Waal for constructive comments on this project and Aiko Hiraguchi and Kaoru Hashimoto for encouragement. Field research was supported by funds from the JSPS International Scientific Research Program (#07041138, #12375003, #16255007, #19255008 to T. Nishida), Global Environment Research Fund of the Ministry of Environment (F-061), the Leakey Foundation, Grant-in-Aid for JSPS Fellows (#15004835 to K. Zamma).

References

Adang OMJ (1984) Teasing in young chimpanzees. Behaviour 88:98–122

Albrecht H, Dunnett SC (1971) Chimpanzees in West Africa. Piper, Munich

Alp R (1997) 'Stepping-sticks' and 'seat-sticks': new types of tools used by wild chimpanzees (*Pan troglodytes*) in Sierra Leone. Am J Primatol 41:45–52

Arcadi AC, Daniel R, Boesch C (1998) Buttress drumming by wild chimpanzees: temporal patterning, phrase integration into loud calls, and preliminary evidence for individual distinctiveness. Primates 39:505–518

Arcadi AC, Wrangham RW (1999) Infanticide in chimpanzees: review of cases and a new within-group observation from the Kanyawara study group in Kibale National Park. Primates 40: 337–351

Assersohn C, Whiten A, Tiwede AT, Tinka J, Karamagi J (2004) Use of leaves to inspect ectoparasites in wild chimpanzees: a third cultural variant? Primates 45:255–258

Azuma S, Toyoshima A (1963) Progress report of the survey of chimpanzees in their natural habitat, Kabogo Point Area, Tanganyika. Primates 3:61–70

Basabose AK, Yamagiwa J (2002) Factors affecting nesting site choice in chimpanzees at Tshibati, Kahuzi-Biega National Park: influence of sympatric gorillas. Int J Primatol 23:263–282

Beck BB (1980) Animal tool behavior. Garland SPTM, New York

Berdecio S, Nash LT (1981) Chimpanzee visual communication. Arizona State University, Tempe, AZ, Anthropological Research Papers No. 26

Bermejo M, Illera G (1999) Tool-set for termite-fishing and honey extraction by wild chimpanzees in the Lossi Forest, Congo. Primates 40:619–627

Bermejo M, Illera G, Sabater Pi J (1994) Animals and mushrooms consumed by bonobos (*Pan paniscus*): New records from Lilungu (Ikela), Zaire. Int J Primatol 15:879–898

Bethell E, Whiten A, Muhumuza G, Kakura J (2000) Active plant food division and sharing by wild chimpanzees. Primate Rep 56:67–71

Biro D, Sousa C, Matsuzawa T (2006) Ontogeny and cultural propagation of tool use by wild chimpanzees at Bossou, Guinea: case studies of nut cracking and leaf folding. In: Matsuzawa T, Tomonaga M, Tanaka M (eds) Cognitive development in chimpanzees. Springer, Tokyo, pp 476–508

Bodenheimer FS (1951) Insects as human food. Dr. W. Junk Publishers, Hague

Boehm C (1999) Hierachy in the forest. Harvard University Press, Cambridge, MA

Boesch C (1991a) The effect of leopard predation on grouping patterns in forest chimpanzees. Behaviour 117:220–242

Boesch C (1991b) Handedness in wild chimpanzees. Int J Primatol 12:541–558

Boesch C (1991c) Teaching in wild chimpanzees. Anim Behav 41:530–532

Boesch C (1995) Innovation in wild chimpanzees (*Pan troglodytes*). Int J Primatol 16:1–16

Boesch C (2009) The real chimpanzee. Cambridge University Press, Cambridge

Boesch C, Boesch H (1990) Tool use and tool making in wild chimpanzees. Folia Primatol 54:86–90

Boesch C, Boesch-Achermann H (2000) The chimpanzees of the Taï Forest. Oxford University Press, Oxford

Boesch C, Head J, Robbins MM (2009) Complex tool sets for honey extraction among chimpanzees in Loango National Park, Gabon. J Hum Evol 56:560–569

Boesch C, Hohmann G, Marchant L (eds) (2002) Behavioural diversity in chimpanzees and bonobos. Cambridge University Press, Cambridge

Bogart SL, Pruetz JD (2008) Ecological context of savanna chimpanzees (*Pan troglodytes verus*) termite fishing at Fongoli, Senegal. Am J Primatol 70:605–612

Bogart SL, Pruetz JD, Kante D (2008) Fongoli chimpanzees (*Pan troglodytes verus*) eats banded mongoose (*Mungos mungo*). Pan Afr News 15:15–17

Bonnie KE, de Waal FBM (2006) Affiliation promotes the transmission of a social custom: handclasp grooming among captive chimpanzees. Primates 47:27–34

Brewer SM, McGrew WC (1990) Chimpanzee use of a tool-set to get honey. Folia Primatol 54:100–104

Bygott JD (1971) Cannibalism among wild chimpanzees. Nature 238:410–411

Bygott JD (1974) Agonistic behaviour and dominance in wild chimpanzees. PhD thesis, University of Cambridge, Cambridge

Byrne R, Whiten A (1988) Machiavellian intelligence. Clarendon, Oxford

Call J, Tomasello M (eds) (2007a) The gestural communication of apes and monkeys. Lawrence Erlbaum Associates, London

Call J, Tomasello M (2007b) The gestural repertoire of chimpanzees (*Pan troglodytes*). In: Call J, Tomasello M (eds) The gestural communication of apes and monkeys. Lawrence Erlbaum Associates, London, pp 17–39

Caro TM, Hauser MD (1992) Is there teaching in nonhuman animals? Q Rev Biol 67:151–174

Chimpanzee Sequencing and Analysis Consortium (2005) Initial sequence of the chimpanzee genome and comparison with the human genome. Nature 437:69–87

Clark AP (1993) Rank differences in the production of vocalizations by wild chimpanzees as a function of social context. Am J Primatol 31:159–179

Constable JL, Ashley MV, Goodall J, Pusey AE (2001) Noninvasive paternity assignment in Gombe chimpanzees. Mol Ecol 10:1279–1300

Corbalis MC (2002) From hand to mouth: The origins of language. Princeton University Press, Princeton

Corp N, Hayaki H, Matsusaka T, Fujita S, Hosaka K, Kutsukake N, Nakamura M, Nakamura M, Nishie H, Shimada M, Zamma K, Wallauer W, Nishida T (2009) Prevalence of muzzle-rubbing and hand-rubbing behavior in wild chimpanzees in Mahale Mountains National Park, Tanzania. Primates 50:184–189

Deblauwe I, Guislain P, Dupain J, van Elsacker L (2006) Use of a tool-set by *Pan troglodytes troglodytes* to obtain termites (*Macrotermes*) in the periphery of the Dja Biosphere Reserve, southeast Cameroon. Am J Primatol 68:1191–1196

Devos C, Gatti S, Levrero F (2002) New record of algae feeding and scooping by *Pan t. troglodytes* at Lokoue Bai in Odzala National Park, Republic of Congo. Pan Afr News 9:19–21

Eibl-Eibesfeldt I (1972) Love and hate. The natural history of behavior patterns. Holt, Rinehart and Winston, New York

Enomoto T (1997) Bonobos. Maruzen, Tokyo (in Japanese)

Fawcett K, Muhumuza G (2000) Death of a wild chimpanzee community member: possible outcome of intense sexual competition? Am J Primatol 51:243–247

Fay JM, Carroll RW (1994) Chimpanzee tool use for honey and termite extraction in Central Africa. Am J Primatol 34:309–317

Feistner ATC, McGrew WC (1989) Food-sharing in primates: a critical review. In: Seth PK, Seth S (eds) Perspectives in primate biology, vol. 3. Today & Tomorrow's, New Delhi, pp 21–36

Fowler A, Sommer V (2007) Subsistence technology of Nigerian chimpanzees. Int J Primatol 28:997–1023

Fruth BI, Hohmann G, Beuerlein MM, McGrew WC (2006) Grooming hand clasp by bonobos of Lui Kotal, Democratic Republic of Congo. Pan Afr News 13:6–8

Fujimoto M, Shimada M (2008) Newly observed predation of wild birds by M group chimpanzees (*Pan troglodytes schweinfurthii*) at Mahale, Tanzania. Pan Afr News 15:23–26

Furuichi T (1989) Social interactions and the life history of female *Pan paniscus* in Wamba, Zaire. Int J Primatol 10:173–197

Furuichi T, Hashimoto C (2000) Ground beds of chimpanzees in the Kalinzu Forest, Uganda. Pan Afr News 7:26–28

Furuichi T, Thompson J (eds) (2008) The bonobos: behavior, ecology, and conservation. Springer, New York

Ghiglieri MP (1984) The chimpanzees of Kibale Forest. Columbia University Press, New York

Gomez J-C, Martin-Andrade B (2005) Fantasy play in apes. In: Pellegrini AD, Smith PK (eds) Thc nature of play. Guilford, New York, pp 139–172

Goodall J (1968) The behavior of free-living chimpanzees in the Gombe Stream Reserve. Anim Behav Monogr 1:161–311

Goodall J (1971) In the shadow of man. Collins, London

Goodall J (1973) Cultural elements in a chimpanzee community. In: Menzel EW (ed) Precultural primate behavior. S. Karger, Basel, pp 144–184

Goodall J (1977) Infant killing and cannibalism in free-living chimpanzees. Folia Primatol 22:259–282

Goodall J (1986) The chimpanzees of Gombe. Harvard University Press, Cambridge, MA

Goodall J (1989) Glossary of chimpanzee behaviors. Jane Goodall Institute, Tucson, AZ

Goodall J (1992) Unusual violence in the overthrow of an alpha male chimpanzee at Gombe. In: Nishida T, McGrew WC, Marler P, Pickford M, de Waal FBM (eds) Topics in primatology 1. Human origins. University of Tokyo Press, Tokyo, pp 131–142

Goodall J, Bandora A, Bergmann E, Busse C, Matama H, Mpongo E, Pierce A, Riss D (1979) Intercommunity interactions in the chimpanzee population of the Gombe National Park. In: The Great Apes, DA Hamburg & ER McCown (eds.), Benjamin/Cummings, Menlo Park, pp.13–53

Graham CE (1981) Menstrual cycle physiology of the great apes. In: Graham CE (ed) Reproductive biology of the great apes. Academic, New York, pp 286–303

Hamai M, Nishida T, Takasaki H, Turner LA (1992) New records of within-group infanticide and cannibalism in wild chimpanzees. Primates 33:151–162

Harris JR (1998) The nurture assumption. Brockman, New York

Hasegawa T (1990) Sex differences in ranging patterns. In: Nishida T (ed) The chimpanzees of the Mahale Mountains. University of Tokyo Press, Tokyo, pp 99–114

Hasegawa T, Hiraiwa-Hasegawa M, Nishida T, Takasaki H (1983) New evidence of scavenging behavior of wild chimpanzees. Curr Anthropol 24:231–232

Hashimoto C, Furuichi T (2005) Possible intergroup killing in chimpanzees in the Kalinzu Forest, Uganda. Pan Afr News 12:3–5

Hashimoto C, Furuichi T (2006) Frequent copulations by females and high promiscuity in chimpanzees in the Kalinzu Forest, Uganda. In: Newton-Fisher NE, Notman H, Paterson JD, Reynolds V (eds) Primates of western Uganda. Springer, New York, pp 247–257

Hashimoto C, Furuichi T, Tashiro Y (2000) Ant dipping and meat eating by wild chimpanzees in the Kalinzu Forest, Uganda. Primates 41:103–108

Hashimoto C, Tashiro Y, Hibino E, Mulabwa M, Yangozene K, Furuichi T, Idani G, Takenaka O (2008) Longitudinal structure of a unit-group of bonobos: male philopatry and possible fusion of unit-groups. In: Furuichi T, Thompson J (eds) The bonobos: behavior, ecology, and conservation. Springer, New York, pp 107–119

Hayaki H (1985a) Social play of juvenile and adolescent chimpanzees in the Mahale Mountains National Park, Tanzania. Primates 26:343–360

Hayaki H (1985b) Copulation of adolescent male chimpanzees, with special reference to the influence of adult males, in the Mahale Mountains National Park, Tanzania. Folia Primatol 44:148–160

Hernandez-Aguilar RA, Moore J, Pickering TR (2007) Savanna chimpanzees use tools to harvest the underground storage organs of plants. Proc Natl Acad Sci USA 104:19210–19213

Hewes GW (1973) Primate communication and the gestural origin of language. Cur Anthropol 14:5–24

Hicks TC, Fouts RS, Fouts DH (2005) Chimpanzee (*Pan troglodytes troglodytes*) tool use in the Ngotto Forest, Central African Republic. Am J Primatol 65:221–237

Hiraiwa-Hasegawa M (1989) Sex differences in behavioral development at Mahale. In: Heltne PG, Marquardt LA (eds) Understanding chimpanzees. Harvard University Press, Cambridge, MA, pp 104–115

Hiraiwa-Hasegawa M, Byrne RW, Takasaki H, Byrne JME (1986) Aggression toward large carnivores by wild chimpanzees of Mahale Mountains National Park, Tanzania. Folia Primatol 47:8–13

Hirata S, Myowa M, Matsuzawa T (1998) Use of leaves as cushions to sit on wet ground by wild chimpanzees. Am J Primatol 44:215–220

Hirata S, Yamakoshi G, Fujita S, Ohashi G, Matsuzawa T (2001) Capturing and toying with hyraxes (*Dendrohyrax dorsalis*) by wild chimpanzees (*Pan troglodytes*) at Bossou, Guinea. Am J Primatol 53:93–97

Hockings KJ, Anderson JR, Matsuzawa T (2010) Flexible feeding on cultivated underground storage organs by rainforest-dwelling chimpanzees at Bossou, West Africa. J Hum Evol 58:227–233

Hockings KJ, Humle T, Anderson JR, Biro D, Sousa C, Ohashi G, Matsuzawa T (2007) Chimpanzees share forbidden fruit. PLoS One 2:e886

Hohmann G (2001) Association and social interactions between strangers and residents in bonobos (*Pan paniscus*). Primates 42:91–99

Hohmann G, Fowler A, Sommer V, Ortmann S (2006) Frugivory and gregariousness of Salonga bonobos and Gashaka chimpanzees: the influence of abundance and nutritional quality of fruit. In: Hohmann G, Robbins MM, Boesch C (eds) Feeding ecology in apes and other primates. Cambridge University Press, Cambridge

Hohmann G, Fruth B (1993) Field observations on meat sharing among bonobos (*Pan paniscus*). Folia Primatol 60: 225-229

Hohmann G, Fruth B (2003) Culture in bonobos? Between-species and within-species variation in behavior. Curr Anthropol 44:563–571

Hohmann G, Fruth B (2007) New records of prey capture and meat eating by bonobos at Lui Kotale, Salonga National Park, Democratic Republic of Congo. Folia Primatol 79:103–110

van Hooff JARAM (1972) A comparative approach to the phylogeny of laughter and smiling. In Hinde RA (ed) Non-verbal communication. Cambridge University Press, Cambridge, pp 362–439

van Hooff JARAM (1973) A structural analysis of the social behavior of a semicaptive group of chimpanzees. In: von Cranach M, Vine I (eds) Social communication and movement. Academic Press, London, pp 75–162

Hopkins WD, Morris RD (1993) Handedness in great apes: A review of findings. Int J Primatol 14:1–25

Hosaka K, Matsumoto-Oda A, Huffman MA, Kawanaka K (2000) Responses of chimpanzees to dead conspecifics. Primate Res 16:1–15

Hosaka K, Nishida T, Hamai M, Matsumoto-Oda A, Uehara S (2001) Predation of mammals by the chimpanzees of the Mahale Mountains, Tanzania. In: Galdikas BMF, Briggs N, Sheeran LK, Shapiro GL, Goodall J (eds) All apes great and small, vol. 1: African apes. Plenum/Kluwer, New York, pp 107–130

Huffman MA (1997) Current evidence for self-medication in primates: a multidisciplinary perspective. Yearb Phys Anthropol 40:171–200

Huffman MA, Kalunde MS (1993) Tool-assisted predation on a squirrel by a female chimpanzee in the Mahale Mountains, Tanzania. Primates 34:93–98

Humle T (1999) New record of fishing for termites (*Macrotermes*) by the chimpanzees of Bossou (*Pan troglodytes verus*), Guinea. Pan Afr News 6:3–4

Humle T, Matsuzawa T (2002) Ant-dipping among the chimpanzees of Bossou, Guinea, and some comparisons with other sites. Am J Primatol 58:133–148

Hunt KD, Cant JG, Gebo DL, Rose MD, Walker SE, Youlatos D (1996) Standardized descriptions of primate locomotor, and postural modes. Primates 37:363–387

Hunt KD, McGrew WC (2002) Chimpanzees in the dry habitats of Assirik, Senegal and Semliki Wildlife Reserve, Uganda. In: Boesch C, Hohmann G, Marchant LF (eds) Behavioural diversity in chimpanzees and bonobos. Cambridge University Press, Cambridge, pp 35–51

Idani G (1990) Relations between unit-groups of bonobos at Wamba, Zaire: encounters and temporary fusions. Afr Study Monogr 11:153–186

Idani G (1991) Social relationships between immigrant and resident bonobo (*Pan paniscus*) females at Wamba. Folia Primatol 57:83–95

Idani G (1995) Function of peering behavior among bonobos (*Pan paniscus*) at Wamba. Primates 36:377–383

Ihobe H (1990) Interspecific interactions between wild pygmy chimpanzees (*Pan paniscus*) and red colobus (*Colobus badius*). Primates 31:109–112

Ihobe H (2005) Life span of chimpanzee beds at the Mahale Mountains National Park, Tanzania. Pan Afr News 12:10–12

Inagaki H, Tsukahara T (1993) A method of identifying chimpanzee hairs in lion feces. Primates 34:109–112

Ingmanson EJ (1996) Tool-using behavior in wild *Pan paniscus*: social and ecological considerations. In: Russon AE, Bard KA, Parker ST (eds) Reaching into thought. Cambridge University Press, Cambridge, pp 190–210

Inoue E, Inoue-Murayama M, Vigilant L, Takenaka O, Nishida T (2008) Relatedness in wild chimpanzees: influence of paternity, male philopatry, and demographic factors. Am J Phys Anthropol 137:256–262

Itani J (1979) Distribution and adaptation of chimpanzees in an arid area. In Hamburg DA, McCown ER (eds) The great apes. Benjamin/Cummings, Menlo Park, CA, pp 55–71

Itoh N, Nishida T (2007) Chimpanzee grouping patterns and food availability in Mahale Mountains National Park, Tanzania. Primates 48:87–96

Itoh N, Sakamaki T, Hamisi M, Kitopeni R, Bunengwa M, Matumla M, Athumani K, Mwami M, Bunengwa H (1999) A new record of invasion into the center of M group territory. Pan Afr News 6:8–10

Iverson JM, Goldin-Meadow S (1998) Why people gesture when they speak. Nature 396:228

Izawa K (1970) Unit-groups of chimpanzees and their nomadism in the savanna woodland. Primates 11:1–46

Jones C, Sabater Pi J (1969) Sticks used by chimpanzees in Rio Muni, west Africa. Nature 223:100–101

Jones C, Sabater Pi J (1971) Comparative ecology of *Gorilla gorilla* (Savage and Wyman) and *Pan troglodytes* (Blumenbach) in Rio Muni, west Africa. S. Karger, Basel

Kahlenberg S, Emery Thompson M, Wrangham RW (2008) Female competition over core areas among *Pan troglodytes schweinfurthii*, Kibale National Park, Uganda. Int J Primatol 29:931–948

Kamenya S (2002) Human baby killed by Gombe chimpanzee. Pan Afr News 9:26

Kano T (1972) Distribution and adaptation of the chimpanzee on the eastern shore of Lake Tanganyika. Kyoto University African Studies 7:37–129

Kano T (1980) Social behavior of wild pygmy chimpanzees (*Pan paniscus*) of Wamba: a preliminary report. J Hum Evol 9:243–260

Kano T (1982a) The social group of pygmy chimpanzees (*Pan paniscus*) of Wamba. Primates 23:171–188

Kano T (1982b) The use of leafy twigs for rain cover by the pygmy chimpanzees of Wamba. Primates 23:453–457

Kano T (1983) An ecological study of the pygmy chimpanzees (*Pan paniscus*) of Yalosidi, Republic of Zaire. Int J Primatol 4:1–31

Kano T (1984) Observations of physical abnormalities among the wild bonobos (*Pan paniscus*) of Wamba, Zaire. Am J Phys Anthropol 63:1–11

Kano T (1990) The bonobo's peaceable kingdom. Natural History, November, pp 62–71

Kano T (1992) The last ape. Stanford University Press, Stanford, CA

Kano T (1998) A preliminary glossary of bonobo behaviors at Wamba. In: Nishida T (ed) Comparative study of the behavior of the genus *Pan* by compiling video ethogram. Report to the Ministry of Education, Culture, Sports, Science and Technology, Nissho Printer, Kyoto, pp 39–81

Kawanaka K (1981) Infanticide and cannibalism in chimpanzees, with special reference to the newly observed case in the Mahale Mountains. Afr Study Monogr 1:69–99

Kiyono M (2008) Use of wet hair to capture swarming termites by a chimpanzee in Mahale, Tanzania. Pan Afr News 15:8–12

Koops K, Humle T, Sterck EHM, Matsuzawa T (2007) Ground-nesting by the chimpanzees of the Nimba Mountains, Guines: environmentally or socially determined? Am J Primatol 69:407–419

Koops K, Matsuzawa T (2006) Hand clapping by a chimpanzee in the Nimba Mountains, Guinea, West Africa. Pan Afr News 13:19–21

Koops K, McGrew WC, Matsuzawa T (2010) Do chimpanzees (*Pan troglodytes*) use cleavers and anvils to fracture *Treculia africana* fruits? Preliminary data on a new form of percussive technology. Primates 51:175–178

Kortlandt A (1962) Chimpanzees in the wild. Sci Am 206:128–138

Kortlandt A (1964) Experimentation with forest-dwelling chimpanzees in the Congo. Verbatim text of 16 mm film

Kortlandt A (1967) Experimentation with chimpanzees in the wild. In: Starck D, Schneider R, Kuhn HJ (eds) Neue Ergebnisse der Primatologie. Fischer, Stuttgart, pp 208–244

Kortlandt A (1996) An epidemic of limb paresis (polio?) among the chimpanzee population at Beni (Zaire) in 1964, possibly tranmitted by humans. Pan Afr News 3(2):9–10

Kortlandt A, Bresser M (1963) Experimentation with forest-dwelling chimpanzees in the Congo, 1963. 8 mm film

Kortlandt A, Kooij M (1963) Protohominid behaviour in primates. Symp Zool Soc Lond 10:61–87

Kummer H, Goodall J (1985) Conditions of innovative behavior in primates. Phil Trans R Soc Lond B 308:441–471

Kuroda S (1979) Grouping of the pygmy chimpanzees. Primates 20:161–183

Kuroda S (1980) Social behavior of the pygmy chimpanzees. Primates 21:181–197

Kuroda S (1984) Interaction over food among pygmy chimpanzees. In: Susman RL (ed) The pygmy chimpanzee. Plenum, New York, pp 301–324

Kuroda S (1998) Preliminary ethogram of tschego chimpanzees. In: Nishida T (ed) Comparative study of the behavior of the genus *Pan* by compiling video ethogram. Report to the Ministry of Education, Culture, Sports, Science and Technology. Nissho Printer, Kyoto, pp 82–91

Kuroda S, Nishihara T, Suzuki S, Oko RA (1996) Sympatric chimpanzees and gorillas in the Ndoki Forest, Congo. In: McGrew WC, Marchant LF, Nishida T (eds) Great ape societies. Cambridge University Press, Cambridge, pp 71–81

Kutsukake N (2006) The context and quality of social relationships affect vigilance behaviour in wild chimpanzees. Ethology 112:581–591

Kutsukake N (2007) Conspecific influences on vigilance behaviour in wild chimpanzees. Int J Primatol 28:907–918

Kutsukake N, Castles DL (2004) Reconciliation and post-conflict third-party affiliation among wild chimpanzees in the Mahale Mountains, Tanzania. Primates 45:157–165

Kutsukake N, Matsusaka T (2002) Incident of intense aggression by chimpanzees against an infant from another group in Mahale Mountains National Park, Tanzania. Am J Primatol 58:175–180

Lanjouw A (2002) Behavioural adaptations to water scarcity in Tongo chimpanzees. In: Boesch C, Hohmann G, Marchant LF (eds) Behavioural diversity in chimpanzees and bonobos. Cambridge University Press, Cambridge, pp 52–60

Lonsdorf EV, Hopkins WD (2005) Wild chimpanzees show population-level handedness for tool use. Proc Natl Acad Sci USA 102:12634–12638

Lycett SJ, Collard M, McGrew WC (2007) Phylogenetic analyses of behavior support existence of culture among wild chimpanzees. Proc Natl Acad Sci USA 104: 17588-17592

Lycett SJ, Collard M, McGrew WC (2009) Cladistic analyses of behavioral variation in wild chimpanzees: exploring the chimpanzee culture hypothesis. J Hum Evol 50:337–349

Manson JH, Wrangham RW (1991) Intergroup aggression in chimpanzees and humans. Curr Anthropol 32:369–390

Marchant LF, McGrew WC (1999) Innovative behavior at Mahale: new data on nasal probe and nipple press. Pan Afr News 6:16–18

Marchant LF, McGrew WC (2007) Ant fishing by wild chimpanzees is not lateralized. Primates 48:22–26
Martin DE, Graham CE, Gould KG (1978) Successful artificial insemination in the chimpanzee. Symp Zool Soc Lond 43:249–260
Matsumoto-Oda A (1997) Self-suckling behavior by a wild chimpanzee. Folia Primatol 68:342–343
Matsumoto-Oda A, Kutsukake N, Hosaka K, Matsusaka T (2006) Sniffing behaviors in Mahale chimpanzees. Primates 48:81–85
Matsusaka T (2004) When does play panting occur during social play in wild chimpanzees? Primates 45:221–229
Matsusaka T, Nishie H, Shimada M, Kutsukake N, Zamma K, Nakamura M, Nishida T (2006) Tool-use for drinking water by immature chimpanzees of Mahale: prevalence of an unessential behavior. Primates 47:113–122
Matsuzawa T (1994) Field experiment on use of stone tools in the wild. In: Wrangham RW, McGrew WC, de Waal FBM, Heltne PG (eds) Chimpanzee cultures. Harvard University Press, Cambridge, MA, pp 351–370
Matsuzawa T (1997) The death of an infant chimpanzee at Bossou, Guinea. Pan Afr News 4:4–6
Matsuzawa T, Biro D, Humle T, Inoue-Nakamura N, Tonooka R, Yamakoshi G (2001) Emergence of culture in wild chimpanzees: education by master-apprenticeship. In: Matsuzawa T (ed) Primate origins of human cognition and behavior. Springer, Tokyo, pp 557–574
Matsuzawa T, Yamakoshi G (1996) Comparison of chimpanzee material culture between Bossou and Nimba, West Africa. In: Russon AE, Bard KA, Parker ST (eds) Reaching into thought. Cambridge University Press, Cambridge, pp 211–232
Matsuzawa T, Yamakoshi G, Humle T (1996) A newly found tool-use by wild chimpanzees: algae scooping (abstract). Primate Res 12:283
McGinnis PR (1973) Patterns of sexual behavior in a community of free-living chimpanzees. PhD thesis, University of Cambridge
McGrew WC (1972) An ethological study of children's behavior. Academic, New York
McGrew WC (1974) Tool use by wild chimpanzees in feeding upon driver ants. J Hum Evol 3:501–508
McGrew WC (1975) Patterns of plant food sharing by wild chimpanzees. In: Kondo S, Kawai M, Ehara A (eds) Contemporary primatology. S Karger, Basel, pp 304–309
McGrew WC (1992) Chimpanzee material culture: implications for human evolution. Cambridge University Press, Cambridge
McGrew WC (1998) Culture in nonhuman primate? Annu Rev Anthropol 27:301–328
McGrew WC (2004) The cultured chimpanzees: reflections on cultural primatology. Cambridge University Press, Cambridge
McGrew WC (2010) In search of the last common ancestor: new findings on wild chimpanzees. Phil Trans R Soc London (in press)
McGrew WC, Baldwin PJ, Marchant LF, Pruetz JD, Scott SE, Tutin CEG (2003) Ethnoarchaeology and elementary technology of unhabituated chimpanzees at Assirik, Senegal, West Africa. Palaeoanthropol 1:1–20
McGrew WC, Baldwin PJ, Tutin CEG (1981) Chimpanzees in a hot, dry and open habitat: Mt. Assirik, Senegal, West Africa. J Human Evol 10:227–244
McGrew WC, Baldwin PJ, Tutin CEG (1988) Diet of wild chimpanzees (*Pan troglodytes verus*) at Mt. Assirik, Senegal. I. Composition. Am J Primatol 16:213–226
McGrew WC, Collins DA (1985) Tool use by wild chimpanzees (*Pan troglodytes*) to obtain termites (*Macrotermes herus*) in the Mahale Mountains, Tanzania. Am J Primatol 9:47–62
McGrew WC, Marchant LF (1998) Chimpanzee wears a knotted skin "necklace." Pan Afr News 5:8–9
McGrew WC, Marchant LF (1999) Laterality of hand use pays off in foraging success for wild chimpanzees. Primates 40:509–513
McGrew WC, Marchant LF, Hunt KD (2007) Ethoarchaeology of manual laterality; well-digging by wild chimpanzees. Folia Primatol 78:240–244

McGrew WC, Marchant LF, Scott SE, Tutin CEG (2001) Intergroup differences in a social custom of wild chimpanzees: the grooming hand-clasp of the Mahale Mountains. Curr Anthropol 42:148–153

McGrew WC, Marchant LF, Wrangham RW, Klein H (1999) Manual laterality in anvil use: wild chimpanzees cracking *Strychnos* fruits. Laterality 4:79–87

McGrew WC, Pruetz JD, Fulton SJ (2005) Chimpanzee use tools to harvest social insects at Fongoli, Senegal. Folia Primatol 76:222–226

McGrew WC, Rogers ME (1983) Chimpanzees, tools and termites: new records from Gabon. Am J Primatol 5:171–174

McGrew WC, Tutin CEG (1978) Evidence for a social custom in wild chimpanzees? Man (n.s.) 13:234–251

McGrew WC, Tutin CEG, Baldwin PJ (1979) Chimpanzees, tools and termites: cross-cultural comparison of Senegal, Tanzania and Rio Muni. Man (n.s.) 14:185–214

Mendoza-Granados D, Sommer V (1995) Play in chimpanzees of the Arnhem Zoo: self-serving compromises. Primates 36:57–68

Mitani JC, Hasegawa T, Gros-Louis J, Marler P, Byrne RW (1992) Dialects in wild chimpanzees? Am J Primatol 27:233–243

Mitani JC, Nishida T (1993) Contexts and social correlates of long-distance calling by male chimpanzees. Anim Behav 45:735–746

Mitani JC, Watts DP (1999) Demographic influence on the hunting behavior of chimpanzees. Am J Phys Anthropol 109:439–454

Mitani JC, Watts DP (2001) Why do chimpanzees hunt and share meat? Anim Behav 61:915–924

Mitani JC, Watts DP (2005) Correlates of territorial boundary patrol behaviour in wild chimpanzees. Anim Behav 70:1079–1086

Mitani JC, Watts DP, Muller MN (2002) Recent developments in the study of wild chimpanzee behavior. Evol Anthropol 11:9–25

Moore J (1996) Savanna chimpanzees, referential models and the last common ancestor. In: McGrew WC, Marchant LF, Nishida T (eds) Great ape societies. Cambridge University Press, Cambridge, pp 275–292

Morgan BJ, Abwe EE (2006) Chimpanzees use stone hammers in Cameroon. Curr Biol 16(16):R632–633

Morgan D, Sanz C (2006) Chimpanzee feeding ecology and comparisons with sympatric gorillas in the Goualougo Triangle, Republic of Congo. In: Hohmann G, Robbins MM, Boesch C (eds) Feeding ecology in apes and other primates. Cambridge University Press, New York, pp 97–122

Mori A (1982) An ethological study on chimpanzees at the artificial feeding place in the Mahale Mountains, Tanzania—with special reference to the booming situation. Primates 23:45–65

Mori A (1983) Comparison of the communicative vocalizations and behaviors of group ranging in eastern gorillas, chimpanzees, and pygmy chimpanzees. Primates 24:486–500

Morin PA, Moore JJ, Chakraborty R, Jin L, Goodall J, Woodruff DS (1994) Kin selection, social structure, gene flow, and the evolution of chimpanzees. Science 265:1193–1201

Morris D (1977) Manwatching: a field guide to human behaviour. Elsevier, Oxford

Morris D (1981) The soccer tribe. Jonathan Cape, London

Morris D (1985) Bodywatching, Equimax, Oxford

Muller MN (2002) Agonistic relations among Kanyawara chimpanzees. In: Boesch C, Hohmann G, Marchant LF (eds) Behavioural diversity in chimpanzees and bonobos. Cambridge University Press, Cambridge, pp 112–124

Muller MN, Mpongo E, Stanford CB, Boehm C (1995) A note on scavenging by wild chimpanzees. Folia Primatol 65:43–47

Nakamura M (1997) First observed case of chimpanzee predation on yellow baboons (*Papio cynocephalus*) at the Mahale Mountains National Park. Pan Afr News 4:9–11

Nakamura M (2003) 'Gatherings' of social grooming among wild chimpanzees: implications for evolution of sociality. J Human Evol 44:59–71

Nakamura M, Itoh N (2001) Sharing of wild fruits among male chimpanzees: two cases of Mahale, Tanzania. Pan Afr News 8:23–31

Nakamura M, Itoh N (2008) Hunting with tools by Mahale chimpanzees. Pan Afr News 15:3–6
Nakamura M, McGrew WC, Marchant LF, Nishida T (1999) Social scratch: another custom in wild chimpanzees? Primates 41:237–248
Nakamura M, Nishida T (2006) Subtle behavior variation in wild chimpanzees, with special reference to Imanishi's concept of *kaluchua*. Primates 47:35–42
Nakamura M, Nishida T (2008) Developmental process of grooming-hand-clasp by chimpanzees of the Mahale Mountains, Tanzania. Primate Eye 96:247
Nakamura M, Uehara S (2004) Proximate factors of different types of grooming-hand-clasp in Mahale chimpanzees: implications for chimpanzee social customs. Curr Anthropol 45:108–114
Newton-Fisher NE (1999) Infant killers of Budongo. Folia Primatol 70:167–169
Nishida T (1968) The social group of wild chimpanzees in the Mahali Mountains. Primates 9:167–224
Nishida T (1970) Social behavior and relationship among wild chimpanzees of the Mahali Mountains. Primates 11:47–87
Nishida T (1973a) The ant-gathering behaviour by the use of tools among wild chimpanzees of the Mahali Mountains. J Hum Evol 2:357–370
Nishida T (1973b) The children of the mountain spirits. Chikuma-shobo, Tokyo (in Japanese)
Nishida T (1976) The bark-eating habits in Primates, with special reference to their status in the diet of wild chimpanzees. Folia Primatol 25:277–287
Nishida T (1979) The social structure of chimpanzees of the Mahale Mountains. In: Hamburg DA, McCown ER (eds) The great apes. Benjamin/Cummings, Menlo Park, CA, pp 72–121
Nishida T (1980a) Local differences in responses to water among wild chimpanzees. Folia Primatol 33:189–209
Nishida T (1980b) The leaf-clipping display: a newly-discovered expressive gesture in wild chimpanzees. J Hum Evol 9:117–128
Nishida T (1983a) Alloparental behavior in wild chimpanzees of the Mahale Mountains, Tanzania. Folia Primatol 41:1–33
Nishida T (1983b) Alpha status and agonistic alliance in wild chimpanzees (*Pan troglodytes schweinfurthii*). Primates 24:318–336
Nishida T (1987) Local traditions and cultural transmission. In: Smuts BB, Cheney DL, Seyfarth RM, Wrangham RW, Struhsaker TT (eds) Primate societies. University of Chicago Press, Chicago, pp 462–474
Nishida T (1989) Social interactions between resident and immigrant female chimpanzees. In: Heltne PG, Marquardt L (eds) Understanding chimpanzees. Harvard University Press, Cambridge, MA, pp 68–89
Nishida T (ed) (1990) The chimpanzee of the Mahale Mountains. University of Tokyo Press, Tokyo
Nishida T (1993a) Left nipple suckling preference in wild chimpanzees. Ethol Sociobiol 14:45–52
Nishida T (1993b) Chimpanzees are always new to me. In: Cavalieri P, Singer P (eds) Great ape project. Fourth Estate, London, pp 24–26
Nishida T (1994) Review of recent findings on Mahale chimpanzees: implications and future research directions. In Wrangham RW, McGrew WC, de Waal FBM, Heltne P (eds) Chimpanzee cultures. Harvard University Press, Cambridge, MA, pp 373–396
Nishida T (1996) The death of Ntologi. The unparalleled leader of M group. Pan Afr News 3(1):4
Nishida T (1997) Sexual behavior of adult male chimpanzees of the Mahale Mountains National Park, Tanzania. Primates 38:379–398
Nishida T (1998) Deceptive tactic by an adult male to snatch a dead infant from a mother. Pan Afr News 5:13–15
Nishida T (2002) A self-medicating attempt to remove the sand flea from a toe by a young chimpanzee. Pan Afr News 9:5–6
Nishida T (2003a) Harassment of mature female chimpanzees by young males in the Mahale Mountains. Int J Primatol 24:503–514

Nishida T (2003b) Individuality and flexibility of cultural behavior patterns in chimpanzees. In: de Waal FBM, Tyack PL (eds) Animal social complexity. Harvard University Press, Cambridge, MA, pp 392–413

Nishida T, Corp N, Hamai M, Hasegawa T, Hiraiwa-Hasegawa M, Hosaka K, Hunt K, Itoh N, Kawanaka K, Matsumoto-Oda A, Mitani JC, Nakamura M, Norikoshi K, Sakamaki T, Turner L, Uehara S, Zamma K (2003) Demography, life history and reproductive profiles among the chimpanzees of Mahale. Am J Primatol 59:99–121

Nishida T, Fujita S, Inaba A, Kooriyama T (2007) Note on a subcutaneous tumor among wild chimpanzees. Pan Afr News 14:31–32

Nishida T, Fujita S, Matsusaka T, Shimada M, Kitopeni R (2007) Dermatophytosis of M group chimpanzees, Mahale Mountains, Tanzania. Pan Afr News 14:5–6

Nishida T, Hasegawa T, Hayaki H, Takahata Y, Uehara S (1992) Meat-sharing as a coalition strategy by an alpha male chimpanzee? In: Nishida T, McGrew WC, Marler P, Pickford M, de Waal FBM (eds) Topics in primatology, vol. 1: Human origins. University of Tokyo Press, Tokyo, pp 159–174

Nishida T, Hiraiwa M (1982) Natural history of a tool-using behavior by wild chimpanzees in feeding upon wood-boring ants. J Hum Evol 11:73–99

Nishida T, Hiraiwa-Hasegawa M (1985) Responses to a stranger mother-son pair in the wild chimpanzee: a case report. Primates 26:1–13

Nishida T, Hiraiwa-Hasegawa M (1987) Chimpanzees and bonobos: cooperative relationships among males. In: Smuts BB, Cheney DL, Seyfarth RM, Wrangham RW, Struhsaker TT (eds) Primate societies. University of Chicago Press, Chicago, pp 165–177

Nishida T, Hiraiwa-Hasegawa M, Hasegawa T, Takahata Y (1985) Group extinction, female transfer in wild chimpanzees in the Mahale National Park, Tanzania. Z Tierpsychol 67:284–301

Nishida T, Hosaka K (1996) Coalition strategies among adult male chimpanzees of the Mahale Mountains, Tanzania. In: McGrew WC, Marchant LF, Nishida T (eds) Great ape societies. Cambridge University Press, Cambridge, pp 114–134

Nishida T, Hosaka K, Nakamura M, Hamai M (1995) A within-group gang attack on a young adult male chimpanzee: ostracism of an ill-mannered member? Primates 36:207–211

Nishida T, Inaba A (2009) Pirouettes: the rotational play of wild chimpanzees. Primates 50:333–341

Nishida T, Kano T, Goodall J, McGrew WC, Nakamura M (1999) Ethogram and ethnography of Mahale chimpanzees. Anthropol Sci 107:141–188

Nishida T, Kawanaka K (1972) Inter-unit-group relationships among wild chimpanzees of the Mahali Mountains. Kyoto Univ Afr Stud 7:131–169

Nishida T, Matsusaka T, McGrew WC (2009) Emergence, propagation or disappearance of novel behavioral patterns in the habituated chimpanzees of Mahale: a review. Primates 50:23–36

Nishida T, Mitani J, Watts DP (2004) Variable grooming behaviours in wild chimpanzees. Folia Primatol 75:31–36

Nishida T, Nakamura M (1993) Chimpanzee tool use to clear a blocked nasal passage. Folia Primatol 61: 218-220

Nishida T, Takasaki H, Takahata Y (1990) Demography and reproductive profiles. In Nishida T (ed) The chimpanzees of the Mahale Mountains. University of Tokyo Press, Tokyo, pp 63–97

Nishida T, Turner LA (1996) Food transfer between mother and infant chimpanzees of the Mahale Mountains National Park, Tanzania. Int J Primatol 17:947–968

Nishida T, Uehara S (1980) Chimpanzees, tools, and termites: another example from Tanzania. Curr Anthropol 21:671–672

Nishida T, Uehara S (1983) Natural diet of chimpanzees (*Pan troglodytes schweinfurthii*): long-term record from the Mahale Mountains, Tanzania. Afr Stud Monogr 3:109–130

Nishida T, Uehara S, Nyundo R (1979) Predatory behavior among wild chimpanzees of the Mahale Mountains. Primates 20:1–20

Nishida T, Wallauer W (2003) Leaf-pile pulling: an unusual play pattern in wild chimpanzees. Am J Primatol 60:167–173

Nishie H (2004) Increased hunting of yellow baboons (*Papio cynocephalus*) by M group chimpanzees at Mahale. Pan Afr News 11:10–12

O'Hara SJ, Lee PC (2006) High frequency of postcoital penis cleaning in Budongo chimpanzees. Folia Primatol 77:353–358

Ohashi G (2006) Behavioral repertoire of tool use in the wild chimpanzees at Bossou. In: Matsuzawa T, Tomonaga M, Tanaka M (eds) Cognitive development in chimpanzees. Springer, Tokyo, pp 439–451

Ohashi G (2007) Papaya fruit sharing in wild chimpanzees at Bossou, Guinea. Pan Afr News 14:14–16

Okayasu N (1991) Vocal communication and its sociological interpretation of wild bonobos in Wamba, Zaire. In: Ehara A, Kimura T, Takenaka O, Iwamoto M (eds) Primatology today. Elsevier, Amsterdam, pp 239–240

Paquette D (1994) Fighting and playfighting in captive adolescent chimpanzees. Aggress Behav 20:49–65

Payne CLR, Webster TH, Hunt KD (2009) Coprophagy by the semi-habituated chimpanzees of Semliki, Uganda. Pan Afr News 15:15–17

Peterson D, Ammann K (2003) Eating apes. University of California Press, Berkeley

Pika S (2007) Gestures in subadult bonobos (*Pan paniscus*). In: Call J, Tomasello M (eds) The gestural communication of apes and monkeys. Lawrence Erlbaum Associates, London, pp 41–67

Pika S, Liebal K, Tomasello M (2005) Gestural communication in subadult bonobos (*Pan paniscus*): repertoire and use. Am J Primatol 65:39–61

Plooij FX (1984) The behavioral development of free-living chimpanzee babies and infants. Ablex, Norwood, NJ

Pollick AS, Jeneson A, de Waal FBM (2008) Gestures and multimodal signalling in bonobos. In Furuichi T, Thompson J (eds) Bonobos: behavior, ecology and conservation. Springer, New York, pp 75–94

Pollick AS, de Waal FBM (2007) Ape gestures and language evolution. Proc Natl Acad Sci USA 104:8184–8189

Pruetz JD (2006) Feeding ecology of savanna chimpanzees (*Pan troglodytes verus*) at Fongoli, Senegal. In: Hohmann G, Robbins MM, Boesch C (eds) Feeding ecology in apes and other primates. Cambridge University Press, New York, pp 161–182

Pruetz JD, Bertolani P (2007) Savanna chimpanzees, *Pan troglodytes verus*, hunt with tools. Curr Biol 17:412–417

Pusey A (1990) Behavioural changes at adolescence in chimpanzees. Behaviour 115:203–246

Pusey A, Murray C, Wallauer W, Wilson M, Wroblewski E, Goodall J (2008) Severe aggression among female *Pan troglodytes schweinfurthii* at Gombe National Park, Tanzania. Int J Primatol 29:949–974

Reynolds V (2005) The chimpanzees of the Budongo Forest. Oxford University Press, Oxford

Reynolds V, Reynolds F (1965) Chimpanzees of the Budongo Forest. In DeVore I (ed) Primate behavior: field studies of monkeys and apes. Holt, Rinehart and Winston, New York, pp 368–424

Sabater Pi J, Bermejo M, Illera G, Vea JJ (1993) Behavior of bonobos (*Pan paniscus*) following their capture of monkeys in Zaire. Int J Primatol 14:797–804

Sakamaki T (1998) First record of algae-feeding by a female chimpanzee at Mahale. Pan Afr News 5:1–3

Sakamaki T (2010) Coprophagy in wild bonobos (*Pan paniscus*) at Wamba in the Democratic Republic of Congo: a possibly adaptive strategy? Primates 51:87–90

Sanz C, Call J, Morgan D (2009) Design complexity in termite-fishing tools of chimpanzees (Pan troglodytes). Biology letters 5:293–296

Sanz CM, Morgan DB (2007) Chimpanzee tool technology in the Goualougo Triangle, Republic of Congo. J Hum Evol 52:420–433

Sanz CM, Morgan DB (2009) Flexible and persistent tool-using strategies in honey-gathering by wild chimpanzees. Int J Primatol 30:411–427

Sanz CM, Morgan DB, Gulick S (2004) New insights into chimpanzees, tools, and termites from the Congo Basin. Am Nat 164:567–581

Schoening C, Ellis D, Fowler A, Sommer V (2007) Army ant prey availability and consumption by chimpanzees (*Pan troglodytes vellerosus*) at Gashaka (Nigeria). J Zool 271:125–133

Schoening C, Humle T, Moebius Y, McGrew WC (2008) The nature of culture: technological variation in chimpanzee predation on driver ants revisited. J Hum Evol 55:48–59

Sherrow HM (2005) Tool use in insect foraging by the chimpanzees of Ngogo, Kibale National Park, Uganda. Am J Primatol 65:377–383

Shimada M (2002) Social scratch among chimpanzees in Gombe. Pan Afr News 9:21–23

Slocombe KE, Newton-Fisher NE (2005) Fruit sharing between wild adult chimpanzees (*Pan troglodytes schweinfurthii*): a socially significant event? Am J Primatol 65:385–391

Stanford CB (1998a) Chimpanzee and red colobus: the ecology of predator and prey. Harvard University Press, Cambridge, MA

Stanford CB (1998b) The social behavior of chimpanzees and bonobos: empirical evidence and shifting assumptions. Curr Anthropol 39:399–420

Stanford CB (2008) Apes of the Impenetrable Forest. Pearson Prentice Hall, Upper Saddle River, NJ

Stanford CB, Gambaneza C, Nkurungi JB, Goldsmith ML (2000) Chimpanzees in Bwindi Impenetrable National Park, Uganda, use different tools to obtain different types of honey. Primates 41:337–341

Stanford CB, Nkurungi JB (2003) Behavioral ecology of sympatric chimpanzees and gorillas in Bwindi Impenetrable National Park, Uganda: diet. Int J Primatol 24:901–918

Stewart FA, Pruetz JD, Hansell MH (2007) Do chimpanzees build comfortable nests? Am J Primatol 69:930–939

Stumpf R (2007) Chimpanzees and bonobos: diversity within and between species. In: Campbell CJ, Fuentes A, Mackinnon KC, Panger M, Bearder SK (eds) Primates in perspective. Oxford University Press, New York, pp 321–344

Sugiyama Y (1968) Social organization of chimpanzees in the Budongo Forest, Uganda. Primates 9:225–258

Sugiyama Y (1981) Observations on the population dynamics and behavior of wild chimpanzees at Bossou, Guinea, in 1979–1980. Primates 22:435–444

Sugiyama Y (1985) The brush-stick of chimpanzees found in south-west Cameroon and their cultural characteristics. Primates 26:361–374

Sugiyama Y (1995a) Tool-use for catching ants by chimpanzees at Bossou and Monts Nimba, West Africa. Primates 36:183–205

Sugiyama Y (1995b) Drinking tools of wild chimpanzees at Bossou. Am J Primatol 37:263–269

Sugiyama Y (1997) Social tradition and the use of tool-composites by wild chimpanzees. Evol Anthropol 6:23–27

Sugiyama Y (1998) Local variation of tool-using repertoire in wild chimpanzees. In: Nishida T (ed) Comparative study of the behavior of the genus *Pan* by compiling video ethogram. Report to the Ministry of Education, Culture, Sports, Science and Technology, Nissho Printer, Kyoto, pp 82–91

Sugiyama Y (2008) The birth of subhuman cultures. Kyoto University Press, Kyoto (in Japanese)

Sugiyama Y, Fushimi T, Sakura O, Matsuzawa T (1993) Hand-preference and use in wild chimpanzees. Primates 34:151–159

Sugiyama Y, Koman J (1987) A preliminary list of chimpanzees' alimentation at Bossou, Guinea. Primates 28:133–147

Surbeck M, Hohmann G (2008) Primate hunting by bonobos of Lui Kotale, Salonga National Park. Curr Biol 18:R906–R907

Susman RL (ed) (1984) The pygmy chimpanzee: evolutionary biology and behavior. Plenum, New York

Susman RL, Badrian NL, Badrian AJ (1980) Locomotor behavior of *Pan paniscus* in Zaire. Am J Phys Anthropol 53:69–80

Suzuki A (1969) An ecological study of chimpanzees in a savanna woodland. Primates 10:103–148

Suzuki A (1971) Carnivority and cannibalism observed among forest-living chimpanzees. J Anthropol Soc Nippon 79:30–48

Suzuki S, Nishihara T, Kuroda S (1995) Tool-set for termite fishing by chimpanzees in the Ndoki Forest, Congo. Behaviour 132:219–235

Takahata Y (1985) Adult male chimpanzees kill and eat a male newborn infant: newly observed intragroup infanticide and cannibalism in Mahale National Park, Tanzania. Folia Primatol 44:161–170

Takahata Y, Hasegawa T, Nishida T (1984) Chimpanzee predation in the Mahale Mountains from August 1979 to May 1982. Int J Primatol 5:213–233

Takahata H, Takahata Y (1989) Inter-unit group transfer of an immature male of the common chimpanzee and his social interactions in the non-natal group. Afr Stud Monogr 9:209–220

Takenoshita Y, Ando C, Iwata Y, Yamagiwa J (2008) Fruit phenology of the great ape habitat in the Moukalaba-Doudou National Park, Gabon. Afr Stud Monogr Suppl 39:23–39

Takenoshita Y, Yamagiwa J, Nishida T (1998) Branch-drop display of a female chimpanzee (*Pan troglodytes troglodytes*) of Petit Loango, Gabon. Pan Afr News 5:16–17

Teleki G (1973) The predatory behavior of wild chimpanzees. Bucknell University Press, Lewisburg, PA

Teleki G (1976) Notes on chimpanzee interactions with small carnivores in Gombe National Park, Tanzania. Primates 14:407–411

Thompson JAM (2002) Bonobos of the Lukuru Wildlife Research Project. In: Boesch C, Hohmann G, Marchant LF (eds) Behavioural diversity in chimpanzees and bonobos. Cambridge University Press, Cambridge, pp 61–70

Tonooka R (2001) Leaf-folding behavior for drinking water by wild chimpanzees (*Pan troglodytes verus*) at Bossou, Guinea. Anim Cogn 4:325–334

Toth N, Schick KD (2009) The Oldowan: the tool making of early hominins and chimpanzees compared. Ann Rev Anthropol 38:289–305

Townsend SW, Slocombe KE, Thompson ME, Zuberbuehler K (2007) Female-led infanticide in wild chimpanzees. Curr Biol 17(10):R356

Tsukahara T (1993) Lions eat chimpanzees: the first evidence of predation by lions on wild chimpanzees. Am J Primatol 29:1–11

Tutin CEG (1975) Exceptions to promiscuity in a feral chimpanzee community. In: Kondo S, Kawai M, and Ehara A (eds) Contemporary primatology. S Karger, Basel, pp 445–449

Tutin CEG (1979a) Mating patterns and reproductive strategies in a community of wild chimpanzees (*Pan troglodytes schweinfurthii*). Behav Ecol Sociobiol 6:29–38

Tutin CEG (1979b) Responses of chimpanzees to copulation, with special reference to interference by immature individuals. Anim Behav 27:845–854

Tutin CEG, Fernandez M, Rogers ME, Williamson EA, McGrew WC (1991) Foraging profiles of sympatric lowland gorillas and chimpanzees in the Lopé Reserve, Gabon. Phil Trans R Soc Lond B 334:178–186

Tutin CEG, McGrew WC, Baldwin PJ (1981) Response of wild chimpanzees to potential predators. In: Chiarelli AB, Corruccini RS (eds) Primate behavior and sociobiology. Springer, Berlin, pp 136–141

Uehara S (1982) Seasonal changes in the techniques employed by wild chimpanzees in the Mahale Mountains, Tanzania, to feed on termites (*Pseudacanthotermes spiniger*). Folia Primatol 37:44–76

Uehara S (1990) Utilization of a marsh grassland within the tropical rain forest by the bonobos (*Pan paniscus*) of Yalosidi, Republic of Zaire. Primates 31:311–322

Uehara S (1997) Predation on mammals by the chimpanzee. Primates 38:193–214

Vigilant L, Hofreiter M, Siedel H, Boesch C (2001) Paternity and relatedness in wild chimpanzee communities. Proc Natl Acad Sci USA 98:12890–12895

de Waal FBM (1982) Chimpanzee politics: power and sex among the apes. Jonathan Cape, London de Waal FBM (1988) The communicative repertoire of captive bonobos (*Pan paniscus*), compared to that of chimpanzees. Behaviour 106:183–251

de Waal FBM (1989) Peace making among primates. Harvard University Press, Cambridge, MA

de Waal FBM (1991) The chimpanzee's sense of social regularity and its relation to the human sense of justice. Am Behav Sci 34:335–349

de Waal FBM (1995) Bonobo sex and society. Sci Am 272:58–64

de Waal FBM (1998) Chimpanzee politics: power and sex among the apes. Revised edition. Johns Hopkins University Press, Baltimore, Maryland

de Waal FBM (ed) (2001) Tree of origin. Harvard University Press, Cambridge, MA

de Waal FBM (2009) The age of empathy. Random House, New York

de Waal FBM, Lanting F (1997) Bonobo. University of California Press, Berkeley, CA

de Waal FBM, Seres M (1997) Propagation of hand clasp grooming among captive chimpanzees. Am J Primatol 43:339–346

de Waal FBM, van Roosmalen A (1979) Reconciliation and consolation among chimpanzees. Behav Ecol Sociobiol 5:55–66

Watts DP (1999) Coalitionary mate-guarding by male chimpanzees at Ngogo, Kibale National Park, Uganda. Behav Ecol Sociobiol 44:43–55

Watts DP (2004) Intracommunity coalitionary killing of an adult male chimpanzee at Ngogo, Kibale National Park, Uganda. Int J Primatol 25:507–521

Watts DP (2008a) Scavenging by chimpanzees at Ngogo and the relevance of chimpanzee scavenging to early hominin behavioral ecology. J Hum Evol 54:125–133

Watts DP (2008b) Tool use by chimpanzees at Ngogo, Kibale National Park, Uganda. Int J Primatol 29:83–94

Watts DP, Mitani JC (2000) Infanticide and cannibalism by male chimpanzees at Ngogo, Kibale National Park, Uganda. Primates 41:357–365

Watts DP, Mitani JC (2001) Boundary patrols and intergroup aggression in wild chimpanzees. Behaviour 138:299–327

Watts DP, Mitani JC (2002) Hunting behavior of chimpanzees at Ngogo, Kibale National Park, Uganda. Int J Primatol 23:1–28

Watts DP, Mitani JC, Sherrow HM (2002) New cases of inter-community infanticide by male chimpanzees at Ngogo, Kibale National Park, Uganda. Primates 43:263–270

Watts DP, Muller M, Amsler SJ, Mbabazi G, Mitani JC (2006) Lethal intergroup aggression by chimpanzees in Kibale National Park, Uganda. Am J Primatol 68:161–180

Webster TH, Hodson P, Hunt KD (2009) Grooming hand-clasp by chimpanzees of the Mugiri community, Toro-Semliki Wildlife Reserve, Uganda. Pan Afr News 16:5–7

Whiten A (1999) Parental encouragement in *Gorilla* in comparative perspective: implications for social cognition and the evolution of teaching. In: Parker ST, Mitchell RW, Miles L (eds) The mentalities of gorillas and orangutans. Cambridge University Press, Cambridge, pp 342–366

Whiten A, Goodall J, McGrew WC, Nishida T, Reynolds V, Sugiyama Y, Tutin CEG, Wrangham RW, Boesch C (1999) Cultures in chimpanzee. Nature 399:682–685

Whiten A, Goodall J, McGrew WC, Nishida T, Reynolds V, Sugiyama Y, Tutin CEG, Wrangham RW, Boesch C (2001) Charting cultural variation in chimpanzees. Behaviour 138: 1481–1516

Williams JM, Pusey AE, Carlis JV, Farm BP, Goodall J (2002) Female competition and male territorial behavior influence female chimpanzees' ranging patterns. Anim Behav 63:347–360

Wilson ML, Wallauer WR, Pusey AE (2004) New cases of intergroup violence among chimpanzees in Gombe National Park, Tanzania. Int J Primatol 25:523–549

Wilson ML, Wrangham RW (2003) Intergroup relationships in chimpanzees. Ann Rev Anthropol 32:363–392

Wittig RM, Boesch C (2003) The choice of post-conflict interactions in wild chimpanzees (*Pan troglodytes*). Behaviour 140:1527–1559

Wrangham RW (1977) Feeding behavior of chimpanzees in Gombe National Park, Tanzania. In: Clutton-Brock TH (ed) Primate ecology. Academic Press, London, pp 503–538

Wrangham RW (1986) Ecology and social relationships in two species of chimpanzees. In: Rubenstein DI, Wrangham RW (eds) Ecological aspects of social evolution: birds and mammals. Princeton University Press, Princeton, NJ, pp 352–378
Wrangham RW (1999) Evolution of coalitionary killing. Yearb Phys Anthropol 42:1–30
Wrangham RW, Chapman CA, Clark-Arcadi AP, Isabirye-Basuta G (1996) Social ecology of Kanyawara chimpanzees: implications for understanding the costs of great ape groups. In: McGrew WC, Marchant LF, Nishida T (eds) Great ape societies. Cambridge University Press, Cambridge, pp 45–57
Wrangham RW, Nishida T (1983) *Aspilia* spp. leaves. A puzzle in the feeding behavior of wild chimpanzees. Primates 24:276–282
Wrangham RW, Peterson D (1996) Demonic males. Houghton Mifflin, Boston
Yamagiwa J, Maruhashi T, Yumoto T, Mwanza N (1996) Dietary and ranging overlap in sympatric gorillas and chimpanzees in the Kahuzi-Biega National Park, Zaire. In: McGrew WC, Marchant LF, Nishida T (eds) Great ape societies. Cambridge University Press, Cambridge, pp 82–98
Yamakoshi G (1998) Dietary responses to fruit scarcity of wild chimpanzees at Bossou, Guinea: possible implications for ecological importance of tool use. Am J Phys Anthropol 106:283–295
Yamakoshi G, Sugiyama Y (1995) Pestle-pounding behavior of wild chimpanzees at Bossou, Guinea: a newly observed tool-using behavior. Primates 36:489–500
Zamma K (2002a) Leaf-grooming by a wild chimpanzee in Mahale. Primates 43:87–90
Zamma K (2002b) A chimpanzee trifling with a squirrel: pleasure derived from teasing? Pan Afr News 9:9–11
Zamma K (2005) Rejecting a bit of meat to get more. Pan Afr News 12:8–10
Zamma K (2006) A louse egg on a leaf. Pan Afr News 13:8–10
Zamma K, Fujita S (2004) Genito-genital rubbing among the chimpanzees of Mahale and Bossou. Pan Afr News 11:5–8

Appendix

List of Copyright Owners of Video Clips

Behavior	**Copyright owner (Video No.)**
abuse	Toshisada Nishida (TN02-441)
abuse carcass	Toshisada Nishida (TN02-235)
adduct penis	Takahisa Matsusaka (MT0197)
aggress	Koichiro Zamma (ZK0497)
aggress in sexual frustration	Toshisada Nishida (TN03-150)
aggress, redirected	Koichiro Zamma (ZK0237)
aid in locomotion	Toshisada Nishida (TN00-126)
	Takahisa Matsusaka (MT0208)
alliance	Toshisada Nishida (TN04-221)
allow	Koichiro Zamma (ZK0141)
alpha male	Toshisada Nishida (TN02-340)
approach	Koichiro Zamma (ZK0335)
approach and withdraw	Toshisada Nishida (TN02-132)
arise	Koichiro Zamma (ZK0273b)
arrive	Toshisada Nishida (TN00-007)
attack	Toshisada Nishida (TN00-149)
attack concertedly	Toshisada Nishida (TN99-023)
avoid	Toshisada Nishida (TN00-233)
avoid fallen log	Toshisada Nishida (TN00-249)
backstroke	Takahisa Matsusaka (MT0016)
balance	Koichiro Zamma (ZK0031)
bark	Toshisada Nishida (TN02-611)
bark, infantile	Takahisa Matsusaka (MT0138)
beat sole with palm	Toshisada Nishida (TN02-138a)
beg	Toshisada Nishida (TN00-231)
bend and stretch knee	Toshisada Nishida (TN03-075)
bend branch	Toshisada Nishida (TN00-015)
bend shrub	Toshisada Nishida (TN02-138b)
bend shrub in courtship	Koichiro Zamma (ZK0160)

bite	Toshisada Nishida (TN00-149)
bite and pull hairs	Toshisada Nishida (TN03-261)
bite self	Takahisa Matsusaka (MT0065)
bob	Toshisada Nishida (TN03-208)
bob bipedal	Toshisada Nishida (TN00-324)
bounce foot/feet up and down rhythmically	Toshisada Nishida (TN00-310)
bow	Toshisada Nishida (TN00-270)
brachiate	Koichiro Zamma (ZK0143a)
break bone at joint	Toshisada Nishida (TN02-229)
break branch	Toshisada Nishida (TN00-186a)
break branch with foot	Toshisada Nishida (TN02-490)
break tree	Toshisada Nishida (TN04-128)
bristle	Koichiro Zamma (ZK0020)
brush away from branch	Toshisada Nishida (TN99-085)
brush away from self	Toshisada Nishida (TN01-124)
brush-tip fishing prob (Goualougo)	By courtesy of Crickette Sanz & David Morgan
care alloparentally	Toshisada Nishida (TN03-307)
care maternally	Toshisada Nishida (TN00-215)
catch with hand	Toshisada Nishida (TN00-063)
change nipple	Toshisada Nishida (TN02-002)
charge	Toshisada Nishida (TN00-002)
chase	Toshisada Nishida (TN00-033)
check penile erection	Toshisada Nishida (TN04-048)
chew	Toshisada Nishida (TN99-025)
circle orthograde	Toshisada Nishida (TN02-099)
circle orthograde off ground	Toshisada Nishida (TN01-144)
circle quadrupedal	Toshisada Nishida (TN02-384)
	Toshisada Nishida (TN01-085)
clack teeth	Toshisada Nishida (TN02-513)
climb	Koichiro Zamma (ZK0077b)
climb cliff	Toshisada Nishida (TN01-146)
climb vertical	Toshisada Nishida (TN99-036)
climb vertical, extended elbow	Toshisada Nishida (TN02-461)
cling	Toshisada Nishida (TN02-343)
clip leaf by hand	Toshisada Nishida (TN02-257)
clip leaf by mouth	Toshisada Nishida (TN02-499)
clip leaf for fishing probe	Toshisada Nishida (TN04-227)
close eyes	Koichiro Zamma (ZK0086)
club	Toshisada Nishida (TN02-278)
club ground	Toshisada Nishida (TN00-235)
club other	Toshisada Nishida (TN00-031)
coalition	Toshisada Nishida (TN04-224)
collect dry leaves	Toshisada Nishida (TN02-087)
compress lips	Takahisa Matsusaka (MT0080)

confiscate	Toshisada Nishida (TN02-613)
console	Takahisa Matsusaka (MT0002)
contact with nipple	Toshisada Nishida (TN03-091)
copulate	Toshisada Nishida (TN00-071)
copulate dorso-ventral	Takahisa Matsusaka (MT0112)
cough	Toshisada Nishida (TN03-329)
cough bark	Toshisada Nishida (TN01-010)
cover nipple	Takahisa Matsusaka (MT0170)
cover self	Toshisada Nishida (TN99-096a)
creep	Takahisa Matsusaka (MT0032)
cross	Koichiro Zamma (ZK0624)
cross arms	Takahisa Matsusaka (MT0080)
cross arms on head	Takahisa Matsusaka (MT0045)
cross legs	Koichiro Zamma (ZK0242)
crouch	Toshisada Nishida (TN02-339)
crowd	Toshisada Nishida (TN04-225)
crush	Koichiro Zamma (ZK0262)
crutch	Koichiro Zamma (ZK0263)
dangle	Takahisa Matsusaka (MT0151)
dart	Koichiro Zamma (ZK0512)
deceive	Toshisada Nishida (TN03-140)
defecate	Koichiro Zamma (ZK0607)
defecate, prone	Koichiro Zamma (ZK0384)
defecate, sitting	Koichiro Zamma (ZK0333b)
depart	Koichiro Zamma (ZK0133)
depart together	Toshisada Nishida (TN00-042)
dermatophytosis	Toshisada Nishida (TN01-100)
descend	Koichiro Zamma (ZK0310)
descend by brachiating	Toshisada Nishida (TN03-305)
descend orthograde feet first	Toshisada Nishida (TN04-201)
descend tree trunk feet first	Toshisada Nishida (TN99-004) Koichiro Zamma (ZK0246)
descend tree trunk head first	Koichiro Zamma (ZK0608)
detain	Toshisada Nishida (TN01-068)
detour	Toshisada Nishida (TN03-226)
dig for underground storage organ by hand	Toshisada Nishida (TN02-009)
dip ant directly (Bossou)	By courtesy of Gaku Ohashi (GO11009)
dip fluid	Takahisa Matsusaka (MT0029)
dip hand and lick water	Toshisada Nishida (TN02-417)
discard	Toshisada Nishida (TN04-097)
discard fruit skin with mouth	Toshisada Nishida (TN03-160)
display as contest	Toshisada Nishida (TN03-083)
display, charging	Toshisada Nishida (TN02-209)
display, rain	Takahisa Matsusaka (MT0205)

display, streambed	Toshisada Nishida (TN99-009)
display toward	Toshisada Nishida (TN00-110)
distract	Toshisada Nishida (TN00-123)
distress call	Toshisada Nishida (TN03-334)
drag and circle	Toshisada Nishida (TN04-145)
drag branch	Toshisada Nishida (TN99-009)
drag carcass by hand	Toshisada Nishida (TN03-024)
drag dry leaves	Toshisada Nishida (TN03-151)
drag other by hand	Toshisada Nishida (TN02-200)
drag other by mouth	Toshisada Nishida (TN02-088)
drag to kill	Agumi Inaba (AI002) Koichiro Zamma (ZK0066)
drape	Koichiro Zamma (ZK0600)
drink	Takahisa Matsusaka (MT0023b)
drink from hole in tree	Takahisa Matsusaka (MT0031)
drink from lake	Toshisada Nishida (TN99-146)
drink from stream	Toshisada Nishida (TN99-001)
drip	Takahisa Matsusaka (MT0093)
drop	Toshisada Nishida (TN03-002)
drop branch	Takahisa Matsusaka (MT0082)
drop infant	Toshisada Nishida (TN00-096)
drop self	Toshisada Nishida (TN02-100)
drum	Toshisada Nishida (TN03-201)
dunk face	Toshisada Nishida (TN00-240)
eat	Koichiro Zamma (ZK0297)
eat algae (Mahale)	By courtesy of Tetsuya Sakamaki
eat algae (Bossou)	By courtesy of Gaku Ohashi (GO10712)
eat beetle larva	Toshisada Nishida (TN99-142)
eat blossom	Toshisada Nishida (TN04-192)
eat *Camponotus* ant	Koichiro Zamma (ZK0620)
eat *Crematogaster* ant	Toshisada Nishida (TN03-106)
eat *Dorylus* ant (Bossou)	By courtesy of Gaku Ohashi (GO11009)
eat egg	Toshisada Nishida (TN03-314)
eat eye mucus	Toshisada Nishida (TN03-104)
eat fruit, inner skin	Toshisada Nishida (TN00-145)
eat fruit, pulp	Toshisada Nishida (TN03-237)
eat gall	Toshisada Nishida (TN03-211)
eat leaf	Toshisada Nishida (TN04-023) Takahisa Matsusaka (MT0184)
eat meat	Toshisada Nishida (TN01-058)
eat nasal mucus	Toshisada Nishida (TN04-065)
eat *Oecophylla* ant	Toshisada Nishida (TN00-207)

eat petiole	Toshisada Nishida (TN03-009)
eat pith	Toshisada Nishida (TN02-453)
eat resin	Toshisada Nishida (TN00-089)
eat root	Toshisada Nishida (TN04-169)
eat seed	Toshisada Nishida (TN04-016)
eat semen	Toshisada Nishida (TN01-097)
eat termite (Mahale)	Takahisa Matsusaka (MT0086)
eat termite (Goualougo)	By courtesy of Crickette Sanz & David Morgan
eat termite soil	Toshisada Nishida (TN02-156)
eat with foot	Toshisada Nishida (TN02-344)
eat xylem	Toshisada Nishida (TN04-169)
ejaculate	Toshisada Nishida (TN00-121)
embrace full	Toshisada Nishida (TN02-016)
embrace half	Koichiro Zamma (ZK0459a)
enter hole	Takahisa Matsusaka (MT0009)
erect penis	Toshisada Nishida (TN00-121)
expel	By courtesy of Michio Nakamura
extend	Toshisada Nishida (TN00-078)
extend arm as ladder	Koichiro Zamma (ZK0139a)
extend hand	Toshisada Nishida (TN00-093)
extend hand, palm downward	Toshisada Nishida (TN02-530)
extend hand, palm sideways	Toshisada Nishida (TN03-192)
extend hand, palm upward	Toshisada Nishida (TN02-483)
extend hand to beg	Koichiro Zamma (ZK0141)
extend leg	Takahisa Matsusaka (MT0043)
extend leg as ladder	Takahisa Matsusaka (MT0208)
fall	Takahisa Matsusaka (MT0084)
fall over backward	Toshisada Nishida (TN00-290)
fart	Koichiro Zamma (ZK0022)
fend	Koichiro Zamma (ZK0141)
fight	Toshisada Nishida (TN00-033)
fill mouth with food	Toshisada Nishida (TN03-159)
fish	Toshisada Nishida (TN99-060)
fish for carpenter ant	Toshisada Nishida (TN04-193)
fish for termite (Goualougo)	By courtesy of Crickette Sanz & David Morgan
flail	Koichiro Zamma (ZK0456)
flail arm	Toshisada Nishida (TN01-041)
flail long object	Toshisada Nishida (TN00-186a)
flee	Takahisa Matsusaka (MT0085)
flee after startle	Toshisada Nishida (TN02-551)
flee from colobus male	Toshisada Nishida (TN03-212)
flip lip	Koichiro Zamma (ZK0041)
flop	Toshisada Nishida (TN02-258)

follow	Toshisada Nishida (TN02-305)
	Toshisada Nishida (TN02-357)
follow in contact	Toshisada Nishida (TN02-545)
follow specific female	Toshisada Nishida (TN04-219)
friendship	Toshisada Nishida (TN03-005)
fumble clitoris	Takahisa Matsusaka (MT0018)
fumble nipple	Toshisada Nishida (TN00-004)
fumble penis	Koichiro Zamma (ZK0127)
fumble penis with foot	Toshisada Nishida (TN03-345)
funny face	Toshisada Nishida (TN04-143)
gallop	Koichiro Zamma (ZK0191)
gaze	Toshisada Nishida (TN02-242)
give	Koichiro Zamma (ZK0610)
glance	Toshisada Nishida (TN00-198)
glove	Toshisada Nishida (TN02-112)
go ahead	Toshisada Nishida (TN02-469)
grab	Toshisada Nishida (TN02-123)
grab and shake	Toshisada Nishida (TN02-425)
grapple	Toshisada Nishida (TN03-046b)
grasp	Toshisada Nishida (TN03-257)
grasp and heave	Toshisada Nishida (TN02-101)
grasp and push shoulders	Toshisada Nishida (TN04-181)
grasp hand	Toshisada Nishida (TN02-572)
greet	Toshisada Nishida (TN02-483)
grin	Toshisada Nishida (TN00-003)
grin-full-closed	Toshisada Nishida (TN02-313)
grin-full-open	Toshisada Nishida (TN99-144)
grin-low-closed	Koichiro Zamma (ZK0255)
grin-low-open	Toshisada Nishida (TN00-004)
groom	Toshisada Nishida (TN02-514)
groom-branch-clasp	Toshisada Nishida (TN99-056)
groom briefly	Toshisada Nishida (TN00-317)
groom by hand	Toshisada Nishida (TN00-181)
groom carcass	Toshisada Nishida (TN02-316)
groom, dyadic	Toshisada Nishida (TN03-146)
groom-hand-clasp	Toshisada Nishida (TN02-602)
groom-hand-clasp unilaterally	Koichiro Zamma (ZK0321)
groom leaf	Koichiro Zamma (ZK0606)
groom mutually	Toshisada Nishida (TN00-054)
groom object or substrate	Toshisada Nishida (TN03-069)
groom, polyadic	Toshisada Nishida (TN04-050)
groom reciprocally	Koichiro Zamma (ZK0601)
groom self	Toshisada Nishida (TN99-067)
groom unilaterally	Toshisada Nishida (TN00-021)

groom with mouth	Toshisada Nishida (TN00-054)
groom wound	Toshisada Nishida (TN00-008)
grope	Toshisada Nishida (TN01-056)
grunt, aha	Toshisada Nishida (TN01-150)
grunt, extended	Takahisa Matsusaka (MT0021)
grunt, food	Toshisada Nishida (TN00-089)
hammer nut with stone (Bossou)	By courtesy of Gaku Ohashi (GO40622)
handicap self	Toshisada Nishida (TN03-132)
hang	Toshisada Nishida (TN00-186b)
hang and spin	Toshisada Nishida (TN00-311)
hang and stamp	Toshisada Nishida (TN00-188)
hang in sloth position	Toshisada Nishida (TN00-057)
hang-stand	Toshisada Nishida (TN02-487)
hang tripedal	Toshisada Nishida (TN01-116)
hang upside-down by feet	Toshisada Nishida (TN02-544)
hang upside-down by hands	Toshisada Nishida (TN99-040)
hang with legs pitterpat	Toshisada Nishida (TN01-006)
hang-wrestle	Toshisada Nishida (TN02-273)
harass	Toshisada Nishida (TN02-404)
headstand	Toshisada Nishida (TN02-302)
herd	Toshisada Nishida (TN03-156)
hesitate	Toshisada Nishida (TN03-066)
hiccup	Toshisada Nishida (TN02-105)
hit	Toshisada Nishida (TN01-007)
hit and run	Toshisada Nishida (TN00-262)
hit bush bipedal	Toshisada Nishida (TN04-007)
hit ground with fist	Toshisada Nishida (TN00-270)
hold body part in mouth	Toshisada Nishida (TN00-197)
hold head or face	Toshisada Nishida (TN01-043)
hold object in groin pocket	Takahisa Matsusaka (MT0013)
hold object in mouth	Toshisada Nishida (TN03-314)
hold object in neck pocket	Takahisa Matsusaka (MT0132)
hold object on head	Takahisa Matsusaka (MT0013)
hoo	Takahisa Matsusaka (MT0185)
hop bipedal on spot	Takahisa Matsusaka (MT0135)
hop quadrupedal on spot	Takahisa Matsusaka (MT0154)
hug	Toshisada Nishida (TN99-127)
hunch and sit	Toshisada Nishida (TN01-095)
hunch bipedal	Toshisada Nishida (TN04-099)
hunch over	Koichiro Zamma (ZK0080)
hunch quadrupedal	Koichiro Zamma (ZK0544)
hunt	Koichiro Zamma (ZK0251)
hunt with tool	By courtesy of Michio Nakamura

hurl self	Toshisada Nishida (TN04-175)
ignore	Toshisada Nishida (TN00-025)
	Koichiro Zamma (ZK0159)
imitate	Toshisada Nishida (TN03-346)
immigrate	Toshisada Nishida (TN02-531)
incest	Toshisada Nishida (TN00-256a)
inspect	Koichiro Zamma (ZK0318)
inspect fruit	Toshisada Nishida (TN04-071)
inspect genitals	Toshisada Nishida (TN00-179)
inspect self	Toshisada Nishida (TN01-021)
inspect wound	Toshisada Nishida (TN00-044)
interfere	Koichiro Zamma (ZK0321)
interfere fishing	Toshisada Nishida (TN04-119)
interfere in copulation	Toshisada Nishida (TN00-183)
interfere play	Takahisa Matsusaka (MT0065)
intervene	Toshisada Nishida (TN02-204)
intervene to separate	Toshisada Nishida (TN00-002)
join play	Toshisada Nishida (TN02-482)
keep water in mouth	Koichiro Zamma (ZK0113a)
kick	Toshisada Nishida (TN02-128)
kick back	Koichiro Zamma (ZK0129)
kick backward quadrupedal	Toshisada Nishida (TN02-424)
kick bipedal	Toshisada Nishida (TN00-312)
kick buttress	Koichiro Zamma (ZK0625)
kick heel	Takahisa Matsusaka (MT0152)
kick other	Toshisada Nishida (TN00-197)
kick up	Toshisada Nishida (TN02-222)
kidnap	Takahisa Matsusaka (MT0144)
kill another species	Toshisada Nishida (TN00-223)
kill time	Koichiro Zamma (ZK0632)
kiss	Toshisada Nishida (TN01-117)
kiss with open mouth	Toshisada Nishida (TN02-484)
kiss with pout face	Takahisa Matsusaka (MT0201)
kiss with tongue	Koichiro Zamma (ZK0614)
knock with both arms	Toshisada Nishida (TN02-225a)
knock with one arm	Toshisada Nishida (TN02-236)
lead	Koichiro Zamma (ZK0209)
leaf-midrib spoon	Toshisada Nishida (TN02-184)
leaf-sponge	Takahisa Matsusaka (MT0031)
leaf-spoon	Toshisada Nishida (TN02-183)
lean	Koichiro Zamma (ZK0425)
leap between trees	Toshisada Nishida (TN01-072)
leap between trees with object	Toshisada Nishida (TN02-225b)
leap bipedal with squared shoulders	Toshisada Nishida (TN03-046a)

leap down	Toshisada Nishida (TN04-081)
leap on	Toshisada Nishida (TN01-111)
leap quadrupedal	Toshisada Nishida (TN02-213)
leap up	Toshisada Nishida (TN04-144)
leap up in surprise	Toshisada Nishida (TN04-014)
leave	Toshisada Nishida (TN99-122)
leave and climb	Toshisada Nishida (TN00-238)
leave to protest	Toshisada Nishida (TN00-245)
lick	Koichiro Zamma (ZK0140)
lick lips	Koichiro Zamma (ZK0113b)
lick rock	Toshisada Nishida (TN99-024)
lick wood	Koichiro Zamma (ZK0034)
lick wound	Toshisada Nishida (TN00-008)
lie	Toshisada Nishida (TN00-155)
lie and hug	Toshisada Nishida (TN02-177)
lie and watch	Toshisada Nishida (TN02-203)
lie lateral	Takahisa Matsusaka (MT0075)
lie on other	Toshisada Nishida (TN02-323)
lie prone	Toshisada Nishida (TN03-221)
lie-sit	Toshisada Nishida (TN04-207)
	Koichiro Zamma (ZK0097a)
lie supine	Toshisada Nishida (TN00-275)
lie supine and shake arms and legs	Toshisada Nishida (TN02-138b)
lie supine with legs apart	Koichiro Zamma (ZK0281)
lie with back to another	Toshisada Nishida (TN02-556)
lie with legs crossed	Koichiro Zamma (ZK0242)
lift	Toshisada Nishida (TN02-518)
lift and drop	Toshisada Nishida (TN03-215)
lift rock	Toshisada Nishida (TN99-024)
limp	Takahisa Matsusaka (MT0125)
listen	Toshisada Nishida (TN01-009)
look around	Toshisada Nishida (TN04-064)
look at water	Takahisa Matsusaka (MT0091)
look back	Koichiro Zamma (ZK0151)
look between thighs	Toshisada Nishida (TN00-018)
look up	Toshisada Nishida (TN02-580)
lower head	Koichiro Zamma (ZK0627)
lower head and shoulder	Koichiro Zamma (ZK0626)
lower leg	Koichiro Zamma (ZK0631)
lower rump	Toshisada Nishida (TN03-019)
make bed	Toshisada Nishida (TN00-017)
make cushion	Toshisada Nishida (TN02-251)
make day bed	Toshisada Nishida (TN00-178)
make ground bed or cushion	Toshisada Nishida (TN02-515)

make night bed	Koichiro Zamma (ZK0621)
make tool	Toshisada Nishida (TN04-195)
massage shoulder	Toshisada Nishida (TN02-290)
misunderstand	Toshisada Nishida (TN00-233)
monitor	Toshisada Nishida (TN00-335)
monitor monkeys	Toshisada Nishida (TN03-092)
monitor mother	Toshisada Nishida (TN00-290)
mop ant	Koichiro Zamma (ZK0193)
mount	Koichiro Zamma (ZK0381)
mount in copulation	Toshisada Nishida (TN00-071)
mount, misdirected	Koichiro Zamma (ZK0079)
mouth	Toshisada Nishida (TN02-471)
mouth for begging	Toshisada Nishida (TN99-125)
move	Toshisada Nishida (TN02-334)
nod and mouth	Toshisada Nishida (TN00-058)
nod to water surface	Toshisada Nishida (TN02-143)
nod to water surface and mouth water	Toshisada Nishida (TN00-029)
nod with body part in mouth	Toshisada Nishida (TN02-180)
nod with object in mouth	Toshisada Nishida (TN02-145)
nod with play face	Toshisada Nishida (TN99-134a)
offer arm	Koichiro Zamma (ZK0134)
offer back	Toshisada Nishida (TN04-006)
open eyes	Koichiro Zamma (ZK0272)
open thighs	Toshisada Nishida (TN00-006)
pant	Koichiro Zamma (ZK0457)
pant-bark	Toshisada Nishida (TN99-104)
pant-grunt	Toshisada Nishida (TN00-314)
pant-grunt with bent elbow	Toshisada Nishida (TN04-106)
pant-hoot	Takahisa Matsusaka (MT0071)
parry	Toshisada Nishida (TN01-107)
pass	Koichiro Zamma (ZK0509)
pass under	Toshisada Nishida (TN02-610)
pass with body contact	Toshisada Nishida (TN03-195)
pat	Toshisada Nishida (TN03-216)
peel with hand	Koichiro Zamma (ZK0370)
peel with teeth	Koichiro Zamma (ZK0062)
peep	Toshisada Nishida (TN04-165)
peer	Toshisada Nishida (TN02-003)
peer together	Toshisada Nishida (TN04-209)
perforate (Goualougo)	By courtesy of Crickette Sanz & David Morgan
pick nose	Takahisa Matsusaka (MT0079)
pick out pulp	Toshisada Nishida (TN03-237)
pick up	Koichiro Zamma (ZK0613)

pick up and release	Toshisada Nishida (TN99-095)
pick up discarded food	Toshisada Nishida (TN00-128)
pile dry leaves	Toshisada Nishida (TN02-399)
pinch clitoris	Takahisa Matsusaka (MT0047)
pirouette	Toshisada Nishida (TN02-379)
play	Koichiro Zamma (ZK0348)
play bite	Koichiro Zamma (ZK0433)
play face	Koichiro Zamma (ZK0340)
play face, full	Koichiro Zamma (ZK0433)
play face, half	Koichiro Zamma (ZK0462)
play in bed	Koichiro Zamma (ZK0196)
play-pant	Toshisada Nishida (TN01-068)
play, parallel	Toshisada Nishida (TN99-149)
play, rough and tumble	Koichiro Zamma (ZK0469)
play socially	Toshisada Nishida (TN02-282)
play socially with object	Takahisa Matsusaka (MT0126)
play solo	Takahisa Matsusaka (MT0121)
play walk	Toshisada Nishida (TN99-096b)
	Koichiro Zamma (ZK0491)
play with object	Takahisa Matsusaka (MT0168)
play with sand	Toshisada Nishida (TN03-089)
play with urine	Takahisa Matsusaka (MT0067)
play with water	Takahisa Matsusaka (MT0055)
poke	Toshisada Nishida (TN02-173)
poke with play face	Toshisada Nishida (TN99-077)
police	Toshisada Nishida (TN01-091b)
	Toshisada Nishida (TN02-021)
pound pestle (Bossou)	By courtesy of Gaku Ohashi (GO10924)
pout	Takahisa Matsusaka (MT0185)
presbyopia	Toshisada Nishida (TN02-067)
present with limbs extended	Koichiro Zamma (ZK0360)
present with limbs flexed	Takahisa Matsusaka (MT0112)
press down	Toshisada Nishida (TN04-147)
press neck with lower arm	Toshisada Nishida (TN99-029)
press object on	Toshisada Nishida (TN02-505)
press teeth against back	Toshisada Nishida (TN02-596)
probe	Koichiro Zamma (ZK0520)
probe nasal passage	Koichiro Zamma (ZK0037)
probe with finger	Toshisada Nishida (TN04-115)
protect	Toshisada Nishida (TN02-345)
protest	Toshisada Nishida (TN02-181)
protrude tongue	Toshisada Nishida (TN02-068)
pucker cheek	Takahisa Matsusaka (MT0198)

pull	Takahisa Matsusaka (MT0159)
pull down	Toshisada Nishida (TN00-206)
pull each other	Takahisa Matsusaka (MT0050)
pull head with hands	Toshisada Nishida (TN99-137)
pull leaf-pile	Toshisada Nishida (TN99-134b)
pull object from opposing sides	Toshisada Nishida (TN01-002)
	Toshisada Nishida (TN02-088)
pull out	Toshisada Nishida (TN03-209)
pull rock to roll	Toshisada Nishida (TN02-164)
pull through with hand	Toshisada Nishida (TN03-340)
pull through with mouth	Toshisada Nishida (TN02-607)
pull with mouth	Toshisada Nishida (TN03-284)
puncture (Goualougo)	By courtesy of Crickette Sanz & David Morgan
push	Toshisada Nishida (TN00-256b)
push ahead	Takahisa Matsusaka (MT0003)
push away	Takahisa Matsusaka (MT0075)
push backward	Takahisa Matsusaka (MT0124)
push down	Toshisada Nishida (TN00-077)
push finger into mouth	Toshisada Nishida (TN00-314)
push forward	Toshisada Nishida (TN03-032)
push head into ventral	Toshisada Nishida (TN03-287)
push leaf-pile	Toshisada Nishida (TN00-339)
push object into	Takahisa Matsusaka (MT0031)
push peri-anogenital region with finger	Koichiro Zamma (ZK0240)
push/pull swing	Toshisada Nishida (TN04-204)
push to shoulder	Toshisada Nishida (TN00-225)
put dorsal	Toshisada Nishida (TN04-032)
put dorsal from ventral	Toshisada Nishida (TN02-472)
put face to	Toshisada Nishida (TN02-234)
put heel on back	Toshisada Nishida (TN04-099)
put mouth into water	Toshisada Nishida (TN00-099)
put rump to rump	Toshisada Nishida (TN00-104)
put ventral	Takahisa Matsusaka (MT0199)
put ventral from dorsal	Takahisa Matsusaka (MT0207)
raise	Koichiro Zamma (ZK0273a)
raise and hold leg	Toshisada Nishida (TN00-334)
raise arm(s) bipedal	Toshisada Nishida (TN04-009)
raise arm quickly	Toshisada Nishida (TN01-062)
raise arm slowly	Koichiro Zamma (ZK0212)
raise arm to hold branch	Toshisada Nishida (TN04-006)
raise arm with elbow bent	Toshisada Nishida (TN03-036)
raise leg while lying	Koichiro Zamma (ZK0328)
raise other's arm	Toshisada Nishida (TN03-189)

raise other's chin	Toshisada Nishida (TN00-051)
raise other's leg	Koichiro Zamma (ZK0630)
rake	Toshisada Nishida (TN04-129)
reach wrist toward	Toshisada Nishida (TN03-240)
reassure	Toshisada Nishida (TN04-106)
rebuff play	Toshisada Nishida (TN00-318)
reconcile	Koichiro Zamma (ZK0614)
regulate direction	Toshisada Nishida (TN03-078)
reingest vomit	Takahisa Matsusaka (MT0040)
reject	Toshisada Nishida (TN01-120)
reject infant	Toshisada Nishida (TN02-475)
reject-move	Koichiro Zamma (ZK0198)
reject-sit	Toshisada Nishida (TN02-073)
relaxed face	Koichiro Zamma (ZK0040)
release infant to fall	Toshisada Nishida (TN03-035)
remove	Koichiro Zamma (ZK0073)
remove lice	Koichiro Zamma (ZK0615)
remove objects from water surface	Toshisada Nishida (TN02-141)
remove sand flea	Toshisada Nishida (TN00-022)
rescue	Toshisada Nishida (TN04-206)
respond to baboon	Toshisada Nishida (TN01-151)
respond to dead animal	Toshisada Nishida (TN99-147)
respond to dead chimpanzee	Koichiro Zamma (ZK0312)
respond to leopard	Toshisada Nishida (TN00-343)
respond to neighboring unit group	Toshisada Nishida (TN02-616)
rest	Koichiro Zamma (ZK0307)
restrain	Toshisada Nishida (TN00-022)
retaliate	Toshisada Nishida (TN00-110)
retreat	Koichiro Zamma (ZK0106)
retreat bipedal	Toshisada Nishida (TN01-041)
retrieve infant	Toshisada Nishida (TN00-215)
reverse	Toshisada Nishida (TN99-102)
ride bipedal	Koichiro Zamma (ZK0467)
ride clinging	Koichiro Zamma (ZK0008)
ride dangling	Takahisa Matsusaka (MT0151)
ride dangling and touch dry leaves	Toshisada Nishida (TN00-212)
ride dorsal	Koichiro Zamma (ZK0097b)
ride jockey	Koichiro Zamma (ZK0208)
ride prone	Koichiro Zamma (ZK0250)
ride quadrupedal	Toshisada Nishida (TN99-141)
	Koichiro Zamma (ZK0072)
ride supine	Koichiro Zamma (ZK0482)
ride ventral	Koichiro Zamma (ZK0021)
ride ventral with limb extended	Toshisada Nishida (TN04-196)

rock back and forth	Toshisada Nishida (TN02-186)
rock side to side	Toshisada Nishida (TN03-082)
roll	Koichiro Zamma (ZK0463)
rotate fruit	Toshisada Nishida (TN02-194)
rub dorsum	Toshisada Nishida (TN01-073)
rub dorsum to conspecific	Toshisada Nishida (TN00-318)
rub genitals	Koichiro Zamma (ZK0204)
rub genitals to substrate	Koichiro Zamma (ZK0100)
rub hand or foot	Takahisa Matsusaka (MT0160)
rub hand with hand	Toshisada Nishida (TN04-098)
rub muzzle	Koichiro Zamma (ZK0453)
rub object to body	Toshisada Nishida (TN00-225)
rummage	Toshisada Nishida (TN04-167)
run	Toshisada Nishida (TN02-528)
run bipedal	Koichiro Zamma (ZK0019)
run quadrupedal	Toshisada Nishida (TN03-127)
scatter dry leaves	Toshisada Nishida (TN02-090)
scoop algae (Bossou)	By courtesy of Gaku Ohashi (GO10712)
scoop infant	Toshisada Nishida (TN04-074)
scramble for food	Toshisada Nishida (TN99-112)
scrape	Koichiro Zamma (ZK0602)
scratch	Koichiro Zamma (ZK0425)
	Takahisa Matsusaka (MT0023a)
scratch dry leaves	Toshisada Nishida (TN03-026)
scratch self	Toshisada Nishida (TN04-106)
scratch self signalling	Koichiro Zamma (ZK0091)
scratch socially	Koichiro Zamma (ZK0069)
scratch socially, poke type (Ngogo)	Toshisada Nishida (TN02-612)
scratch socially, stroke type	Toshisada Nishida (TN00-251)
scream	Toshisada Nishida (TN03-280)
search for conspecific	Takahisa Matsusaka (MT0026)
search for object	Toshisada Nishida (TN03-074)
senescence	Toshisada Nishida (TN01-045)
separate	Toshisada Nishida (TN03-304)
shake arm, abduct	Toshisada Nishida (TN02-423)
shake arm, adduct	Takahisa Matsusaka (MT0091)
shake back and forth with four limbs	Toshisada Nishida (TN02-287)
shake branch	Toshisada Nishida (TN00-010)
shake branch up and down with feet	Toshisada Nishida (TN02-036)
shake face side to side	Toshisada Nishida (TN00-240)
shake face with object in mouth	Toshisada Nishida (TN02-426)
shake hand side to side quickly	Toshisada Nishida (TN02-115)
shake head	Takahisa Matsusaka (MT0010)

shake long object irregularly	Toshisada Nishida (TN02-245)
shake object up and down	Toshisada Nishida (TN02-500)
shake off	Toshisada Nishida (TN04-132)
shake other	Toshisada Nishida (TN02-571)
shake penis	Koichiro Zamma (ZK0136)
shake rock up and down with feet	Toshisada Nishida (TN00-026)
shake rump	Toshisada Nishida (TN03-163)
share food	Toshisada Nishida (TN01-058)
shrug	Toshisada Nishida (TN03-163)
sickness	Toshisada Nishida (TN03-339)
sit	Koichiro Zamma (ZK0623)
	Koichiro Zamma (ZK0633)
sit and turn back	Koichiro Zamma (ZK0603)
sit behind another	Koichiro Zamma (ZK0459b)
sit face to face	Koichiro Zamma (ZK0628)
sit on	Toshisada Nishida (TN01-070)
sit prone	Koichiro Zamma (ZK0629)
sit sideways	Koichiro Zamma (ZK0328)
slap	Koichiro Zamma (ZK0126)
slap buttress or tree trunk	Toshisada Nishida (TN02-214)
slap ground	Toshisada Nishida (TN01-091a)
slap in invitation	Toshisada Nishida (TN99-075)
slap other	Toshisada Nishida (TN00-041)
slap self	Toshisada Nishida (TN03-264)
	Toshisada Nishida (TN03-253)
slap-stamp	Koichiro Zamma (ZK0077a)
slap wall	Toshisada Nishida (TN02-043)
sleep	Koichiro Zamma (ZK0256)
slide down boulder	Takahisa Matsusaka (MT0107)
slide down vertically	Toshisada Nishida (TN01-110)
smack lip	Toshisada Nishida (TN01-042)
snatch	Toshisada Nishida (TN00-129)
sneer	Takahisa Matsusaka (MT0113)
sneeze	Koichiro Zamma (ZK0265)
sniff	Toshisada Nishida (TN00-216)
sniff finger	Koichiro Zamma (ZK0387)
sniff fruit	Toshisada Nishida (TN99-054)
sniff mouth	Toshisada Nishida (TN02-255)
sniff with tool	Koichiro Zamma (ZK0520)
snub	Toshisada Nishida (TN01-131)
solicit companion	Agumi Inaba (AI009)
solicit copulation	Toshisada Nishida (TN02-506)
solicit grooming	Toshisada Nishida (TN04-006)
solicit grooming turn	Koichiro Zamma (ZK0601)

solicit play	Toshisada Nishida (TN02-202)
	Toshisada Nishida (TN00-018)
solicit play with object	Toshisada Nishida (TN02-210)
solicit play with object in mouth	Toshisada Nishida (TN02-239)
solicit reassurance contact	Koichiro Zamma (ZK0237)
solicit riding	Takahisa Matsusaka (MT0202)
solicit support	Toshisada Nishida (TN03-045)
somersault	Koichiro Zamma (ZK0103)
somersault, backward	Toshisada Nishida (TN02-512)
somersault, backward with dry leaves	Toshisada Nishida (TN02-260)
somersault, forward	Toshisada Nishida (TN03-020)
somersault, side	Toshisada Nishida (TN02-357)
spit juice	Toshisada Nishida (TN02-588)
spit seed	Toshisada Nishida (TN02-308)
spit water	Koichiro Zamma (ZK0214)
splash water	Takahisa Matsusaka (MT0131)
spring	Toshisada Nishida (TN02-434)
sputter (Ngogo)	Toshisada Nishida (TN03-347)
squash leaf	Koichiro Zamma (ZK0606)
squat	Toshisada Nishida (TN02-328)
squeal in copulation	Koichiro Zamma (ZK0110)
squeeze	Toshisada Nishida (TN00-074)
staccato call	Takahisa Matsusaka (MT0173)
stamp	Toshisada Nishida (TN01-006)
stamp bipedal	Toshisada Nishida (TN01-013)
stamp in invitation	Toshisada Nishida (TN02-033)
stamp other	Takahisa Matsusaka (MT0076)
stamp quadrupedal	Toshisada Nishida (TN00-038)
stamp trot	Takahisa Matsusaka (MT0119)
stamp water	Toshisada Nishida (TN02-147)
stand bipedal	Toshisada Nishida (TN00-185)
stand quadrupedal	Koichiro Zamma (ZK0616)
stand quadrupedal heel up	Koichiro Zamma (ZK0622)
stand with head down, bottom up	Toshisada Nishida (TN04-186)
stare fixedly	Toshisada Nishida (TN00-233)
stare fixedly with head down and bottom up	Toshisada Nishida (TN03-066)
step on	Toshisada Nishida (TN00-061)
step over	Takahisa Matsusaka (MT0200)
step up on leg	Toshisada Nishida (TN01-048)
stir water	Toshisada Nishida (TN99-002)
	Takahisa Matsusaka (MT0054)
store	Toshisada Nishida (TN00-143)
strip leaf	Toshisada Nishida (TN00-071)
stroke	Toshisada Nishida (TN02-422)
stumble	Takahisa Matsusaka (MT0049)

subcutaneous tumor	Toshisada Nishida (TN00-340)
suck	Toshisada Nishida (TN02-002)
suck in reassurance	Toshisada Nishida (TN02-402)
suck self	Koichiro Zamma (ZK0063)
suck thumb	Takahisa Matsusaka (MT0088)
	Takahisa Matsusaka (MT0044)
suck toe	Toshisada Nishida (TN02-136)
suckle	Koichiro Zamma (ZK0324)
supplant	Toshisada Nishida (TN99-063)
support	Koichiro Zamma (ZK0217)
support dominant	Toshisada Nishida (TN03-280)
support older	Toshisada Nishida (TN03-294)
support subordinate	Toshisada Nishida (TN03-087)
suspend	Toshisada Nishida (TN02-223)
suspend and shake up and down	Toshisada Nishida (TN02-224)
swagger bipedal	Toshisada Nishida (TN01-062)
swagger on knuckles	Toshisada Nishida (TN00-006)
swallow leaf	Toshisada Nishida (TN03-048)
sway and move	Takahisa Matsusaka (MT0042)
sway woody vegetation	Toshisada Nishida (TN02-599)
	Takahisa Matsusaka (MT0195)
swing	Toshisada Nishida (TN03-302)
swing above	Toshisada Nishida (TN02-215)
swing and grasp	Toshisada Nishida (TN02-568)
swing and kick	Toshisada Nishida (TN02-010)
swing forward and upward	Toshisada Nishida (TN02-122)
take	Koichiro Zamma (ZK0141)
	Koichiro Zamma (ZK0051)
take finger in mouth	Koichiro Zamma (ZK0143b)
tap heel	Toshisada Nishida (TN02-447)
tear	Toshisada Nishida (TN04-226)
tease	Toshisada Nishida (TN00-049)
threaten	Toshisada Nishida (TN02-484)
throw	Toshisada Nishida (TN00-223)
throw at animate object	Toshisada Nishida (TN02-153)
throw at inanimate object	Toshisada Nishida (TN02-191)
throw branch	Koichiro Zamma (ZK0020)
throw dry leaves	Koichiro Zamma (ZK0222)
throw sand	Toshisada Nishida (TN03-165)
throw splash	Toshisada Nishida (TN03-276)
	Koichiro Zamma (ZK0337)
throw stone or rock	Koichiro Zamma (ZK0349)
throw temper tantrum	Toshisada Nishida (TN99-123)
	Toshisada Nishida (TN00-245)
thrust	Takahisa Matsusaka (MT0070)

thrust bipedal	Toshisada Nishida (TN00-226)
thrust in vacuum	Toshisada Nishida (TN02-035)
thrust, misdirected	Toshisada Nishida (TN99-144)
tickle	Koichiro Zamma (ZK0317)
tilt head	Toshisada Nishida (TN99-100)
tool set (Goualougo)	By courtesy of Crickette Sanz & David Morgan
touch	Toshisada Nishida (TN04-106)
touch fruit	Toshisada Nishida (TN04-071)
touch scrotum	Toshisada Nishida (TN01-141)
touch with foot	Toshisada Nishida (TN04-041)
toy	Koichiro Zamma (ZK0187)
trample	Takahisa Matsusaka (MT0076)
transfer	Takahisa Matsusaka (MT0007)
transport	Koichiro Zamma (ZK0618)
transport bipedal	Toshisada Nishida (TN03-330)
transport corpse of infant	Takahisa Matsusaka (MT0012)
transport food	Koichiro Zamma (ZK0257)
transport in foot	Toshisada Nishida (TN03-278)
transport in groin pocket	Toshisada Nishida (TN03-002)
transport in hand	Koichiro Zamma (ZK0139b)
transport in mouth	Toshisada Nishida (TN02-237) Toshisada Nishida (TN02-614)
transport in neck pocket	Toshisada Nishida (TN02-510)
transport on back	Toshisada Nishida (TN02-042)
transport on head or nape	Toshisada Nishida (TN04-156)
transport on shoulder	Toshisada Nishida (TN02-232)
transport two offspring	Toshisada Nishida (TN04-205)
transport with hand support	Koichiro Zamma (ZK0619)
travel	Toshisada Nishida (TN02-286) Koichiro Zamma (ZK0507)
travel and play	Toshisada Nishida (TN00-142)
trifle with	Koichiro Zamma (ZK0067)
tumble	Takahisa Matsusaka (MT0109)
turn around	Koichiro Zamma (ZK0335)
turn face away	Toshisada Nishida (TN04-214)
turn up lip	Toshisada Nishida (TN01-045)
twist	Toshisada Nishida (TN02-230)
urinate	Takahisa Matsusaka (MT0067)
urinate, prone	Toshisada Nishida (TN03-224)
urinate quadrupedal	Takahisa Matsusaka (MT0133)
urinate, sitting	Koichiro Zamma (ZK0333a)
use tool (Bossou)	By courtesy of Gaku Ohashi (GO40622 and GO10712)

use tool (Goualougo)	By courtesy of Crickette Sanz & David Morgan
use tool (Mahale)	Takahisa Matsusaka (MT0031)
	Koichiro Zamma (ZK0124)
vacate	Toshisada Nishida (TN03-190)
vomit	Toshisada Nishida (TN99-098)
waa bark	Toshisada Nishida (TN00-336)
wade	Toshisada Nishida (TN02-459)
wadge	Toshisada Nishida (TN01-133)
	Koichiro Zamma (ZK0114)
wadge by adding leaf	Toshisada Nishida (TN00-039)
wadge without adding leaf	Toshisada Nishida (TN01-115)
wait for companion	Koichiro Zamma (ZK0480)
wait turn	Koichiro Zamma (ZK0050)
walk bipedal	Toshisada Nishida (TN02-455)
walk in sloth position	Koichiro Zamma (ZK0154)
walk quadrupedal on backs of hands	Takahisa Matsusaka (MT0107)
walk quadrupedal on knuckles	Koichiro Zamma (ZK0053)
walk quadrupedal on palms	Takahisa Matsusaka (MT0107)
	Koichiro Zamma (ZK0283)
walk tripedal	Toshisada Nishida (TN00-320)
watch	Koichiro Zamma (ZK0275)
wean	Takahisa Matsusaka (MT0170)
wedge	Toshisada Nishida (TN02-595)
whimper	Koichiro Zamma (ZK0107)
whimper-scream	Takahisa Matsusaka (MT0005)
whisk fly with arm	Koichiro Zamma (ZK0313)
wipe with detached object	Koichiro Zamma (ZK0003)
wraa	Toshisada Nishida (TN00-242)
wrestle	Toshisada Nishida (TN03-132)
wrestle bipedal	Toshisada Nishida (TN99-100)
wrestle with fingers	Takahisa Matsusaka (MT0078)
wriggle	Toshisada Nishida (TN00-265)
yawn	Toshisada Nishida (TN03-312)
	Koichiro Zamma (ZK0244)

Printed in Japan

Springer